SPIDER MITES

of

SOUTHWESTERN UNITED STATES

and a revision of the family

TETRANYCHIDAE

DONALD M. TUTTLE   and   EDWARD W. BAKER
University of Arizona   U.S. Department of
Agriculture

THE UNIVERSITY OF ARIZONA PRESS
Tucson, Arizona

CONTENTS

REVISION OF THE FAMILY TETRANYCHIDAE
(Tuttle and Baker 1966)

SUBFAMILY BRYOBIINAE BERLESE

TRIBE BRYOBIINI RECK
Genus *Bryobia* Koch
Genus *Parabryobia,* new genus
Genus *Bryobiella,* new genus

TRIBE HYSTRICHONYCHINI PRITCHARD AND BAKER
Genus *Hystrichonychus* McGregor
Genus *Tetranycopsis* Canestrini
Genus *Porcupinychus* Anwarullah
Genus *Monoceronychus* McGregor
Genus *Reckia* Wainstein
Genus *Mesobryobia* Wainstein
Genus *Parapetrobia* Meyer and Ryke
Genus *Aplonobia* Womersley
Genus *Paraplonobia* Wainstein, new status
    Subgenus *Paraplonobia* Wainstein
    Subgenus *Langella* Wainstein, new status
Genus *Anaplonobia* Wainstein
Genus *Neopetrobia* Wainstein
Genus *Beerella* Wainstein
Genus *Georgiobia* Wainstein
Genus *Schizonobiella* Beer and Lang

TRIBE PETROBIINI RECK
Genus *Petrobia* Murray
    Subgenus *Petrobia* Murray
    Subgenus *Tetranychina* Banks
    Subgenus *Mesotetranychus* Reck
Genus *Schizonobia* Womersley
Genus *Mezranobia* Athias-Henriot

TRIBE NEOTRICHOBIINI, NEW TRIBE
Genus *Neotrichobia,* new genus

SUBFAMILY TETRANYCHINAE BERLESE

TRIBE EURYTETRANYCHINI RECK
Genus *Eurytetranychus* Oudemans
Genus *Eutetranychus* Banks
Genus *Aponychus* Rimando
Genus *Synonychus* Miller

TRIBE TENUIPALPOIDINI PRITCHARD AND BAKER
Genus *Tenuipalpoides* Reck and Bagdasarian

TRIBE TETRANYCHINI RECK
Genus *Panonychus* Yokoyama
Genus *Allonychus* Pritchard and Baker
Genus *Eotetranychus* Oudemans
Genus *Schizotetranychus* Trägårdh
Genus *Neotetranychus* Trägårdh
Genus *Mononychus* Wainstein
Genus *Platytetranychus* Oudemans
Genus *Anatetranychus* Womersley
Genus *Tylonychus* Miller
Genus *Mixonychus* Ryke and Meyer
Genus *Oligonychus* Berlese
    Subgenus *Oligonychus* Berlese
    Subgenus *Wainsteiniella,* new subgenus
    Subgenus *Homonychus* Wainstein
    Subgenus *Metatetranychoides* Wainstein
    Subgenus *Reckiella,* new subgenus
    Subgenus *Pritchardinychus* Wainstein
Genus *Tetranychus* Dufour
    Subgenus *Tetranychus* Dufour
    Subgenus *Polynychus* Wainstein
    Subgenus *Armenychus* Wainstein

v

INTRODUCTION

Interest in the spider mites is growing throughout the world, and since the mid 1950s several monographic papers have been published. The major works are those by Pritchard and Baker (1955), Bagdasarian (1957), Reck (1959), and Wainstein (1960). Other works have been published, and those by Ehara on the Tetranychidae of Japan are especially important.

In a previous paper, Tuttle and Baker (1964) published on the tetranychids of Arizona. Since then, the senior author has made extensive collections throughout the state, finding many new species and genera. This has given us an opportunity to re-evaluate the systematics of the family as proposed by Pritchard and Baker (1955) and Wainstein (1960).

By using claw and empodial and striation patterns, we have somewhat changed the relationships set up by the preceding authors. Most genera and subgenera have been kept, although at times changed in rank. We have not here accepted Wainstein's tribes Monoceronychini and Beerellini. We have also added a new tribe, Neotrichobiini, in the Bryobiinae.

Holotypes are deposited in the U. S. National Museum.

Recent studies on chromosomes in several genera of the tetranychids made by Dr. W. Helle of Holland (in press) indicate that those genera with similar empodia are more closely related than those with dissimilar empodia. The genera included in these studies were *Bryobia, Tetranycopsis, Neotetranychus, Schizotetranychus, Oligonychus,* and *Tetranychus.*

Grateful acknowledgement is made to Dr. Charles T. Mason, Jr., and Mr. Wesley E. Niles, Department of Botany, University of Arizona, for identification of many of the plants appearing in this publication.

Donald M. Tuttle
Edward W. Baker

TETRANYCHIDAE DONNADIEU

Tétranycidés Donnadieu, 1875: 9.
Tetranychidae, Murray, 1877: 93, 97; Pritchard and Baker, 1955: 4;
Wainstein, 1960: 88.

The Tetranychidae possess long recurved whiplike movable chelae set in the stylophore or fused basal segments of the chelicerae; the fourth papal segment bears a strong claw; the tarsi I and II, and sometimes the tibiae, usually bear specialized duplex setae; the claws possess tenent hairs, and the empodium may or may not have tenent hairs; the female genitalia is characteristically wrinkled, and the male genitalia is characteristic of the family as well as of the species. Normally there are three pairs of propodosomal setae, four pairs of marginal setae, and one pair of humeral setae. Setae may shift, drop out, or extra pairs may be added.

The family as accepted here possesses two subfamilies, the Bryobiinae and the Tetranychinae.

KEY TO SUBFAMILIES OF TETRANYCHIDAE

1. Empodium with tenent hairs; female with three pairs of anal setae and male with five pairs of genito-anal setae . . . . . . Bryobiinae 2
2. Empodium absent or if present without tenent hairs; female with two pairs of anal setae and male with four pairs of genito-anal setae . . .
 . . . . Tetranychinae 80

1

BRYOBIINAE BERLESE

Bryobiini Berlese, 1913: 17.
Bryobiinae Reck, 1950: 122; Pritchard and Baker, 1955: 12;
  Wainstein, 1960: 91.

The Bryobiinae has four tribes, the Bryobiini, Hystrichonychini, Petrobiini, and Neotrichobiini.

KEY TO THE TRIBES AND GENERA OF BRYOBIINAE (FEMALES)

1. True claws uncinate; empodium padlike (Bryobiini) 2
   True claws padlike; empodium padlike or uncinate 4
2. With three pairs of propodosomal setae 3
   With four pairs of propodosomal setae . . . . . . *Bryobia* Koch
3. Tarsus I with normal two sets of duplex setae; para-anal setae ventral . .
   . . . . *Parabryobia,* new genus
   Tarsus I without duplex setae; para-anal setae dorsal . . . .
   . . . . *Bryobiella,* new genus
4. Claws and empodium padlike (Hystrichonychini) 5
   Claws padlike and empodium uncinate 18
5. With three pairs of propodosomal setae 6
   With four pairs of propodosomal setae . . . *Tetranycopsis* Canestrini
6. With ten or more pairs of hysterosomal setae 7
   With eight pairs of hysterosomal setae; all setae strong and set on strong
   tubercles . . . . *Porcupinychus* Anwarullah
7. Fourth pair of dorsocentral setae marginal 8
   Fourth pair of dorsocentral setae in normal position or nearly so . . . 11
8. Dorsocentral setae not set on strong tubercles 9
   Dorsocentral setae on strong tubercles . . . . *Beerella* Wainstein
9. With propodosomal projections over rostrum 10
   Without projections over rostrum; without dorsal shields; integument with
   tuberculate pattern . . . . *Reckia* Wainstein
10. With two anterior projections over rostrum . *Mesobryobia* Wainstein
    With three anterior projections over rostrum, *Monoceronychus* McGregor
11. With ten pairs of hysterosomal setae (including humerals) 12
    With twelve pairs of hysterosomal setae . . *Hystrichonychus* McGregor

12. Female with normal two sets of duplex setae on tarsus I 13
    Female with three sets of duplex setae on tarsus I . . . .
    . . . . *Parapetrobia* Meyer and Ryke

2

3

BYROBIINI RECK

Bryobiinae Reck, 1952: 423.
Bryobiini, Pritchard and Baker, 1955: 14; Wainstein, 1960: 93.

This tribe is characterized by having the true claws uncinate and the empodium padlike. The following genera are included here:

Bryobia Koch
Parabryobia, new genus
Bryobiella, new genus

Bryobia Koch

Bryobia Koch, 1836: 1: 8, 9; Pritchard and Baker, 1955: 14;
Wainstein, 1960: 94.
Pseudobryobia McGregor, 1950: 355; Pritchard and Baker, 1955: 15;
Wainstein, 1960: 113.
Type. Bryobia praetiosa Koch.

Mites of the genus Bryobia have four pairs of propodosomal setae. The true claws are uncinate and possess tenent hairs; the empodium is padlike and has tenent hairs. There may or may not be propodosomal projections over the rostrum. The peritremes are either simple or anastomose distally.

KEY TO THE SPECIES OF BRYOBIA IN ARIZONA (FEMALES)

1. With strong anterior propodosomal projections 2
   Without such projections 5
2. True claws of leg I strongly hooked 3
   True claws of leg I straight, not hooked . convolvulus Tuttle and Baker
3. Stylophore rounded anteriorly; dorsal striae not strong 4
   Stylophore deeply cleft; dorsal striae strong . . . curiosa Summers
4. Duplex setae of tarsi III and IV nearly of equal length . . . . . .
   . . . . rubrioculus (Scheuten)
   Solenidion of duplex setae of tarsi III and IV about twice as long as
   tactile seta . . . . praetiosa Koch
5. Setae not set on strong tubercles 6
   Setae long, slender and set on strong tubercles . namae Tuttle and Baker

4

6.   Setae slender; stylophore not indented; leg I about as long as body . .   7
      Setae squamiform; stylophore indented anteriorly; leg I about one and
      a half times as long as body . . . . *ephedrae,* new species
7.   Solenidion of tarsus IV subequal in length to dorsal tactile setae . . .
    . . . . *filifoliae,* new species
      Solenidion of tarsus IV one-half as long as dorsal tactile setae . . . .
    . . . . *drummondi* (Ewing)

Bryobia convolvulus Tuttle and Baker

Bryobia convolvulus Tuttle and Baker, 1964: 6.

This species has propodosomal projections over the rostrum. The peritreme ends in a simple bulb. The tarsal claws of leg I are not hooked as are those of the other legs, consisting only of a small pointed spur. The dorsal body setae are slightly serrated. Tarsus III has a short proximal solenidion separated from the much longer (3x) tactile seta; The tactile seta of tarsus IV is proximal and subequal in length to the adjacent solenidion.

This species was collected on *Convolvulus arvensis* L. in Arizona.

Bryobia curiosa Summers

Bryobia curiosa Summers, 1953: 290.

This species lacks the strong anterior propodosomal projections, all true claws are hooked, and the stylophore is deeply cleft.

Bryobia curiosa was originally found on composite desert plants in southern California. It has been collected in Arizona on *Antennaria arida* E. Nels., McNary, August 22, 1964, by D. M. T. Another collection was made from *Ephedra californica* Wats., Alpine, California, July 26, 1963, by D. M. T.

Bryobia rubrioculus (Scheuten)

Sannio rubrioculus Scheuten, 1857: 104.
Bryobia rubrioculus, Tuttle and Baker, 1964: 6.

Bryobia rubrioculus, the fruit tree species, has long been confused with *B. praetiosa* Koch, the grass species. *B. rubrioculus* is also known in literature as *B. arborea* Morgan and Anderson (1957).

Bryobia rubrioculus has propodosomal projections over the rostrum. The peritremes are anastomosing distally. The solenidion and proximal tactile seta on tarsus III are approximate and subequal in length; these setae are reversed in position on tarsus IV, are subequal in length, and are separated. This mite overwinters on the trees in the egg stage.

This species has been taken on apple (*Malus sylvestris* Mill.) in Arizona.

Bryobia praetiosa Koch

Bryobia praetiosa Koch, 1836: 8; Pritchard and Baker, 1955: 26;
Wainstein, 1960: 102; Tuttle and Baker, 1964: 4.

Bryobia praetiosa, which infests grasses and low-lying plants, is similar
to *B. rubrioculus,* differing in the setation of tarsi III and IV. The tactile seta
on tarsus III is proximal and approximate to the solenidion, and about half
its length; the same pattern is found on tarsus IV. This mite overwinters
as an adult.

Bryobia praetiosa is widely distributed in Arizona, especially in the agri-
cultural areas. It infests many grasses and weeds and can also be a nuisance
by invading houses in the winter and spring. Arizona host records include:
Aegilops cylindrica Host., *Arctium minus* Schuhr., *Brassica juncea* (L.) Cosson,
Brassica sp., *Calendula officinalis* L., *Cenchrus echinatus* L., *Centaurea cyanus*
L., *Centaurea imperialis* Hausskn, *Chenopodium album* L., *Chenopodium
murale* L., *Chrysanthemum* sp., *Cortaderia selloana* (Schult.) Ashers &
Graebn., *Cucumis sativus* L., *Cucurbita* sp., *Cucurbita pepo* L., *Heterotheca
subaxillaris* (Lam.) Britt. & Rusby, *Hordeum arizonicum* Covas, *Ipomoea*
sp., Lichen, *Lolium perenne* L., *Malva* sp., *Malva parviflora* L., *Medicago
hispida* Gaertn., *Medicago sativa* L., *Melilotus indica* (L.), *Muhlenbergia
pulcherrima* Scribn., *Plantago insularis* Eastw., *Portulaca oleracea* L., *Prunus
amygdalus* Batsch, *Pteridium aquilinum* (L.) Kuhn, *Pyrus communis* L.,
Raphanus sativus L., *Rumex crispus* L., *Secale cereale* L., *Spergularia* sp.,
Tagetes sp., *Trifolium dubium* Sibth., and *Trifolium repens* L.

Bryobia namae Tuttle and Baker

Bryobia namae Tuttle and Baker, 1964: 6.

This species lacks the propodosomal projections over the rostrum. The
peritremes anastomose distally. The dorsal body setae are long, strong and
serrate, and, except for the anterior two pairs of propodosomal setae, are
set on strong tubercles. The propodosoma is covered with large tubercles;
the hysterosoma has a few crenulate transverse striae dorsally. The solenidia
of tarsi III and IV are short, lanceolate, distal and not associated with tactile
setae.

Bryobia namae has been taken on *Nama hispidum* Gray in Arizona.

6

Bryobia ephedrae, new species

(Figs. 1-4)

This species is characterized by the lack of propodosomal projections and in having the propodosoma cleft between the anterior pair of setae. The stylophore is strongly cleft. The body setae are ovate and serrate and are not set on tubercles. The solenidion on tarsus IV is slender, distal, and is not associated with a tactile seta.

Female. Rostrum small; femoral setae short, lanceolate-serrate. Stylophore deeply cleft; peritremes anastomosing distally. Propodosoma without projections over rostrum; anterior two pairs of setae set on small tubercles, inner pair smaller than outer; propodosoma cleft between inner pair of setae; all propodosomal setae broadly rounded and serrate. Propodosoma covered with large tubercles; hysterosoma with few irregular transverse striae. Hysterosomal setae broadly clavate and serrate. Leg I very long and slender, at least one and one-half times as long as body; duplex setae of leg I with very short and long members; empodium with a single pair of tenent hairs; claws strong and curved. Empodium and claws of other legs similar. Duplex setae of tarsus II short and subequal in length. Tarsus III and IV each with a slender short solenidion not associated with a tactile seta. All dorsal leg setae strong, lanceolate serrate. Length of body 626 μ; including rostrum 720 μ.

Male. Not known.

Holotype. Female, *ex Ephedra trifurca* Torr., Portal, Arizona, August 28, 1964, by D. M. T.

Paratype. A female with the above data.

7

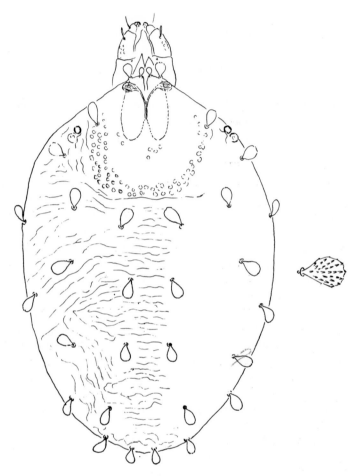

Figure 1, Dorsum of female.
Bryobia ephedrae, new species

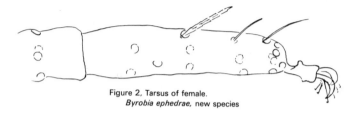

Figure 2, Tarsus of female.
Byrobia ephedrae, new species

Figure 3, Propodosomal projections of female.
Bryobia ephedrae, new species

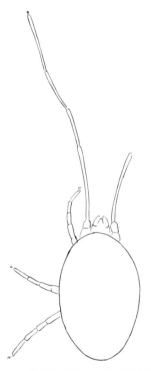

Figure 4, Body and legs of female.
Bryobia ephedrae, new species

Bryobia filifoliae, new species
(Figs. 5-6)

This species lacks the propodosomal projections, the dorsal body setae are rodlike, serrate or slightly expanded distally, and the duplex setae of tarsus IV are subequal in length.

Female. Rostrum small, broad; femoral setae short, lanceolate-serrate. Stylophore rounded distally; peritremes anastomosing distally. Propodosoma without projections over rostrum; anterior two pairs of setae not set on tubercles, serrate and slightly expanded distally, the first pair about two-thirds as long as second pair; both pairs serrate and slightly expanded distally; second pair about as long as other propodosomal setae. Propodosoma lightly punctate dorsomedially; hysterosoma with few weak transverse striae. Hystersosmal setae similar to posterior propodosomal setae, serrate as figured. Leg I not as long as body; tarsus I with two pairs of duplex setae with members of unequal size, and with two solenidia; tibia I with slender distal solenidion. Tarsus II with one set of duplex setae of unequal size and one slender solenidion; tibia II without solenidion. Tarsi III and IV similar, each with a single dorsal solenidion. Dorsal leg setae strong, lanceolate-serrate. Empodia all short and each with a pair of tenent hairs; claws strong and curved, with tenent hairs. Length of body 541 μ; including rostrum 613 μ.

Male. Not known.

Holotype. Female, *ex Artemisia filifolia* Torr., McNary, Arizona, August 22, 1964, by D. M. T.

Paratype. A female, with the above data.

Bryobia drummondi (Ewing)

Petrobia drummondi Ewing, 1926: 143.
Bryobia drummondi, Pritchard and Baker, 1955: 19; Tuttle and Baker, 1964: 4.

Bryobia drummondi lacks the propodosomal projections over the rostrum; the peritremes anastomose distally; the first pair of propodosomal setae are less than one-half as long as the second pair; and the solenidion of tarsus IV is about one-half as long as the tactile seta anterior to it.

This species has been collected in Arizona on *Cucurbita palmata* Wats., *Ephedra fasciculata* A. Nels., *Larrea tridentata* (DC.) Coville, and *Sorghum halepense* (L.) Pers.

10

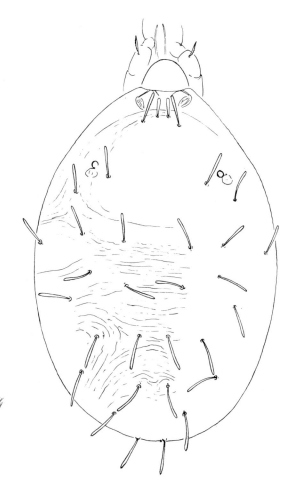

Figure 5, Dorsum of female.
Bryobia filifoliae, new species

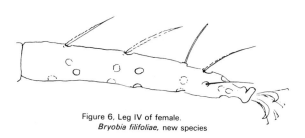

Figure 6, Leg IV of female.
Bryobia filifoliae, new species

Parabryobia, new genus

Type. Parabryobia deleoni, new species.

This genus is distinctive in having only three pairs of propodosomal setae; the claws are hooked and the empodia are padlike.

Female with only three pairs of propodosomal setae; with twelve pairs of hysterosomal setae; tarsal claws uncinate and with tenent hairs; empodia padlike and with tenent hairs; tarsus I with two sets of duplex setae; Tarsus II without duplex setae but with small solenidion; Tarsi III and IV each with solenidion and tactile seta as in *Bryobia;* dorsal body setae arranged as in *Bryobia* except for anterior median propodosomal setae which are not present; peritreme anastomosing distally.

Parabryobia deleoni, new species

(Figs. 7-11)

This species of distinctive in having only three pairs of propodosomal setae, and in having uncinate claws and padlike empodia as in *Bryobia.*

Female. Rostrum small; palpal femoral seta large, lanceolate-serrate; stylophore broadly rounded anteriorly; peritreme anastomosing and slender distally. Propodosoma without anterior projections over rostrum, with only three pairs of setae, all clavate, serrate and longer than wide; striae as figured. Hysterosoma with twelve pairs of dorsal setae (including humeral setae), arranged as in *Bryobia,* the fourth pair of dorsocentral setae marginal or nearly marginal; all setae broadly clavate and with few weak serrations; body with few strong transverse striae. All legs shorter than body; all tarsi with uncinate claws and padlike empodium; empodium I with a single pair of tenent hairs; other empodia with three pairs of tenent hairs; tarsus I with two sets of duplex setae and one lateral solenidion; tarsus II with a short distal solenidion but no duplex setae; dorsal setae slender on tarsus and tibia, but broadly clavate on other segments. Tarsi III and IV each with a short solenidion, and a much longer distal tactile seta. Length of body 440 μ; including rostrum 480 μ.

Male. Not known.

Holotype. Female, *ex Selaginella* plants, Bustamante, Nuevo Leon, Mexico, at Laredo, Texas Quarantine, June 3, 1965, by Kodama and Stang.

Paratypes. Six females with the above data.

Although not from Arizona, this species is from the general area, and it is included here to complete the discussion on the genera of the Bryobiinae.

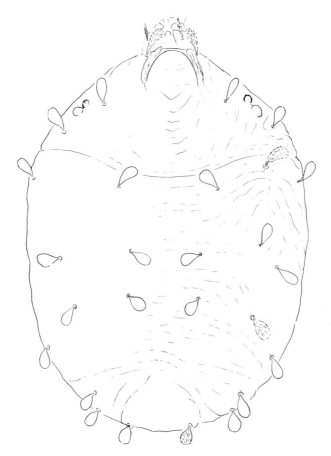

Figure 7, Dorsum of female.
Parabryobia deleoni, new species

13

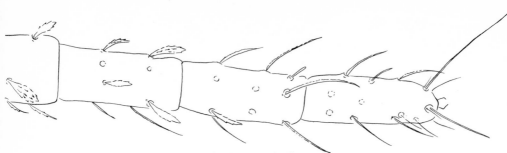

Figure 8, Leg I of female.
Parabryobia deleoni, new species

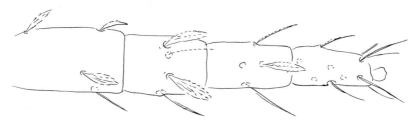

Figure 9, Leg II of female.
Parabryobia deleoni, new species

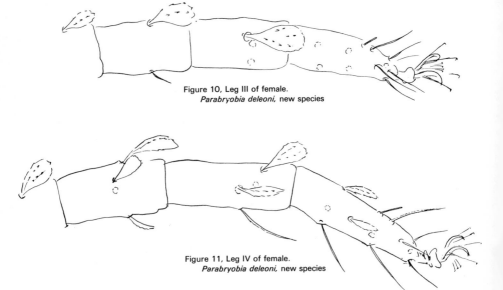

Figure 10, Leg III of female.
Parabryobia deleoni, new species

Figure 11, Leg IV of female.
Parabryobia deleoni, new species

14

Bryobiella, new genus

Type. Bryobiella desertorum, new species.

This genus is distinctive in having the para-anal setae lying dorsally; in lacking the propodosomal projections over the rostrum; in having the true claws uncinate and the empodia padlike; and in lacking duplex setae on tarsus I.

With only three pairs of propodosomal setae; with fourteen pairs of hysterosomal setae — this includes the humeral setae and the two pairs of para-anal setae which lie on the posterior dorsal section of the body just anterior to the terminal anal opening; true claws uncinate and each with a single pair of tenent hairs; tarsus I without duplex setae but with two solenidia in female and six solenidia in male; tarsus II without duplex setae but with a single solenidion; no solenidion on legs III or IV; peritreme ending in simple bulb.

Bryobiella desertorum, new species

(Figs. 12-21)

The arrangement of the para-anal setae and the lack of duplex setae are distinctive for this species.

Female. Propodosoma with three pairs of setae; hysterosoma with fourteen pairs of setae including humeral and para-anal setae; striae longitudinal on propodosoma and transverse on hysterosoma. Peritreme ending in simple elongate bulb; stylophore narrow and rounded anteriorly. Palpal setae small and simple; dorsal body setae slightly lanceolate and serrate; anal setae forked. Tarsus I without duplex setae, with two solenidia; tibia I with five tactile setae and one solenidion; genu I with five tactile setae; femur I with four tactile setae. Tarsus II with one solenidion and no duplex setae; tibia II with five tactile setae; genu II with five tactile setae; femur II with four tactile setae. Tarsus III with eight tactile setae; tibia III with five tactile setae; genu III with four tactile setae; femur III with two tactile setae. Tarsus and tibia IV similar to III; genu IV with three tactile setae; femur IV with one tactile seta. Leg setae slender and pilose. No solenidia on legs III and IV. True claws hooked and each with a pair of tenent hairs; empodium padlike and with a pair of tenent hairs. Length of body 466 μ; including rostrum 600 μ.

Male. Setal and striation patterns similar to those of female. Aedeagus bryobiid-like in being long, strong and slightly curved. Tarsus I without duplex setae but with six solenidia; tibia I with five tactile setae and seven solenidia; genu I with five tactile setae and four solenidia; femur I with four tactile setae and four solenidia. Tarsus II with one solenidion; tibia II with five tactile setae and one solenidion; genu II with five tactile setae and two solenidia; femur II with four tactile setae and one solenidion. Tarsus III with

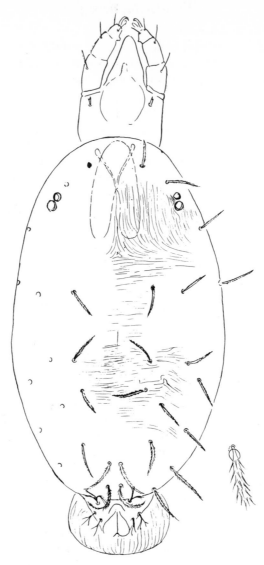

Figure 12, Dorsum of female.
Bryobiella desertorum, new species

eight tactile setae; tibia III with five tactile setae; genu III with four tactile
setae; femur III with two tactile setae. Tarsus and tibia IV similar to III;
genu IV with three tactile setae; femur IV without tactile setae. Claws and
empodia as in female. Length of body 333 μ; including rostrum 426 μ.

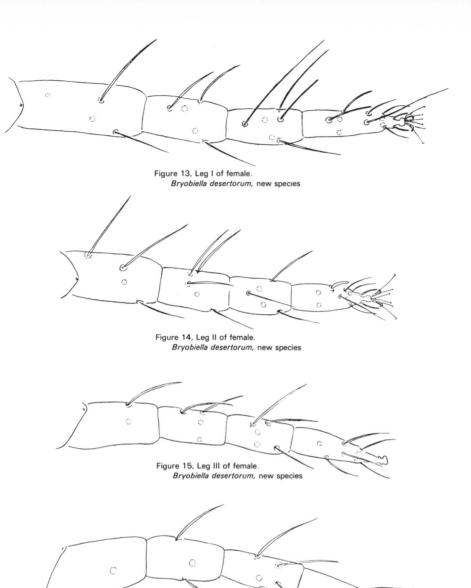

Figure 13, Leg I of female.
Bryobiella desertorum, new species

Figure 14, Leg II of female.
Bryobiella desertorum, new species

Figure 15, Leg III of female.
Bryobiella desertorum, new species

Figure 16, Leg IV of female.
Bryobiella desertorum, new species

Holotype. Female, *ex Euphorbia albomarginata* Torr. and Gray, Gila Bend, Arizona, October 15, 1963, by D. M. T.

Paratype. Male, *ex Euphorbia polycarpa* Benth., Gila Bend, Arizona, April 9, 1963, by D. M. T.

17

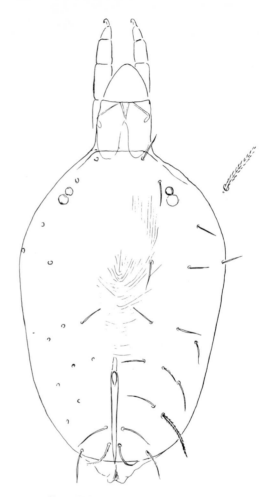

Figure 17, Dorsum of male.
Bryobiella desertorum, new species

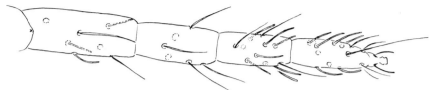

Figure 18, Leg I of male.
Bryobiella desertorum, new species

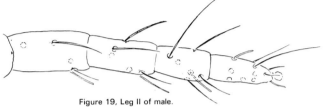

Figure 19, Leg II of male.
Bryobiella desertorum, new species

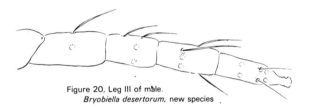

Figure 20, Leg III of male.
Bryobiella desertorum, new species .

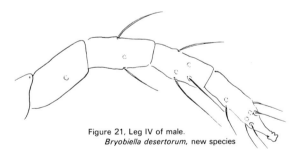

Figure 21, Leg IV of male.
Bryobiella desertorum, new species

HYSTRICHONYCHINI PRITCHARD AND BAKER

Hystrichonychini Pritchard and Baker, 1955: 35; Wainstein, 1960: 120.

This tribe is characterized by having both claws and empodia padlike. The following genera are included here:

Hystrichonychus McGregor
Tetranycopsis Canestrini
Porcupinychus Anwarullah
Monoceronychus McGregor
Reckia Wainstein
Mesobryobia Wainstein
Parapetrobia Meyer and Ryke

Aplonobia Womersley
Paraplonobia Wainstein
Anaplonobia Wainstein
Neopetrobia Wainstein
Beerella Wainstein
Georgiobia Wainstein
Schizonobiella Beer and Lang

Hystrichonychus McGregor

Hystrichonychus McGregor, 1950: 272; Pritchard and Baker, 1955: 37; Wainstein, 1960: 121.
Neotetranycopsis Bagdasarian, 1951: 370.

Type. *Tetranychus gracilipes* Banks.

There are three pairs of propodosomal setae, twelve pairs of hysterosomal setae plus one pair of humeral setae, all long and strong and set on prominent tubercles. The peritreme is anastomosing distally. The true claws are short and padlike with a pair of terminal tenent hairs. The male is similar to the female except in having the first pair of propodosomal setae placed far anteriorly.

KEY TO THE SPECIES OF HYSTRICHONYCHUS IN ARIZONA (FEMALES)
1. First pair of anterior propodosomal setae on transverse line with second pair 2
   First pair of propodosomal setae on line posterior to second pair . . . . *spinosus,* new species
2. Second pair of propodosomal setae much shorter than third pair . . 3
   Second pair of propodosomal setae equal in length to third pair . . . . *sidae* Pritchard and Baker
3. Femur I with three setae; distal seta not reaching end of segment . . . . *ambiguae,* new species
   Femur I with four or five setae; distal seta reaching to or surpassing end of segment . . . . *gracilipes* (Banks)

Hystrichonychus spinosus, new species

(Figs. 22-26)

This species is distinctive in having long, strong dorsal body setae, and in that the first pair of propodosomal setae are set on a transverse line well behind the second pair, which are much shorter than either the first or third pair of propodosomal setae.

Female. Anterior pair of propodosomal setae in line behind and inside second pair of setae; second pair of propodosomal setae about half as long as third pair; first pair longest. Hysterosomal setae long, strong, subequal in length except for the shorter fifth pair of dorsocentral setae; setal serrations fine; all setae, especially on hysterosoma, set on well defined tubercles. Peritremes anastomosing distally; stylophore rounded or slightly indentate anteriorly; rostrum not as long as in *Hystrichonychus gracilipes*. Leg setation as figured; femoral and genual setae relatively long and strong; distal femoral I seta surpassing end of segment; dorsal genual seta also surpassing end of segment. Femur I with large distal seta and five small setae, the three proximal setae serrate and stronger than others. Length of body 280 μ; including rostrum 373 μ.

Male. Not known.

Holotype. Female, *ex Sphaeralcea ambigua* Gray, Portal, Arizona, August 28, 1964, by D. M. T.

Paratypes. Three females with the above data.

Hystrichonychus sidae Pritchard and Baker

Hystrichonychus sidae Pritchard and Baker, 1955: 40; Tuttle and Baker, 1964: 8.

This species is distinctive in having the first pair of propodosomal setae almost in line with the second pair; the second and third pairs of setae are subequal in length, and the first pair is slightly longer. Other body setae longer and subequal in length, but not as long as in *Hystrichonychus gracilipes*. The serrations of the setae are weak. The leg setae are short and stout. The male has short body setae; the first pair of propodosomal setae is on the anterior margin of the propodosoma. The aedeagus is short, strong, and upturned but not sigmoid. The peritreme is slightly branching distally.

Hystrichonychus sidae has been collected on *Sida hederacea* (Dougl.) Torr. in California. It has been collected in Arizona on *Encelia farinosa* Gray, *Sphaeralcea ambigua* Gray, and on *Sphaeralcea* sp. in Sonora, Mexico.

21

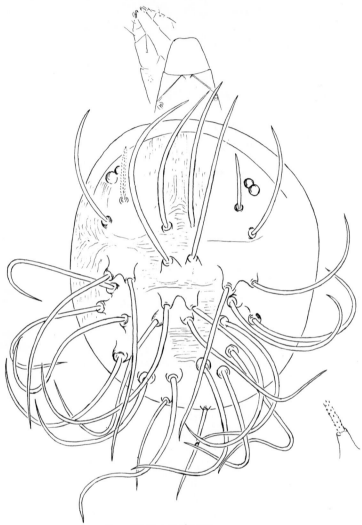

Figure 22, Dorsum of female.
Hystrichonychus spinosus, new species

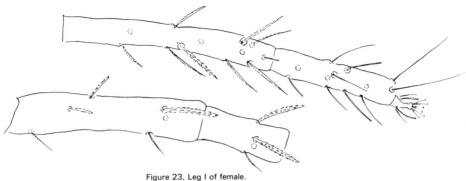

Figure 23, Leg I of female.
Hystrichonychus spinosus, new species

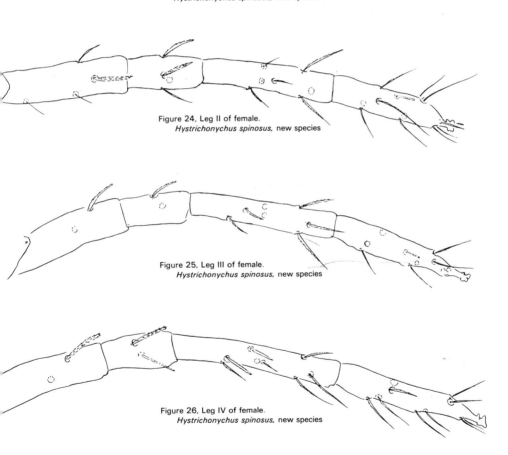

Figure 24, Leg II of female.
Hystrichonychus spinosus, new species

Figure 25, Leg III of female.
Hystrichonychus spinosus, new species

Figure 26, Leg IV of female.
Hystrichonychus spinosus, new species

Hystrichonychus ambiguae, new species

(Figs. 27-32)

This species is distinctive in that the first pair of propodosomal setae are on a transverse line with the short second pair, and in that the distal femoral I seta does not reach the end of the segment. The male is similar to that of *Hystrichonychus gracilipes,* and, at present, the two males cannot be separated.

Female. Anterior pair of propodosomal setae inside and on a transverse line with second pair; second pair of setae short, not half as long as first; all other dorsal body setae as long as or longer than first pair of propodosomal setae. All setae, especially hysterosomal setae, set on prominent tubercles; serrations of setae small and few. Peritremes slightly branching distally. Stylophore not elongate, only slightly indentate anteriorly; rostrum not as long as in *Hystrichonychus gracilipes.* Leg setation as figured; femur I with short dorsodistal broadly lanceolate seta which does not reach end of segment; with three to four short slender setae; and with a single long, strong proximal ventral seta; genu I seta short, broad and strong. Length of body 320 μ; including rostrum 413 μ.

Male. First pair of propodosomal setae on anterior portion of propodosoma; first three pairs of propodosomal setae subequal in length; all body setae short in comparison with those of female; serrations of setae stronger than in female; setal tubercles not strong. Leg setae stronger than in female. Aedeagus short, strong, upturned and slightly sigmoid distally. Length of body 230 μ; including rostrum 293 μ.

Holotype. Female, *ex Sphaeralcea ambigua* Gray, Amarillo, Texas, June 17, 1964, by D. M. T.

Paratype. Male with the above data.

It has also been collected on *Sphaeralcea coccinea* (Pursh), Rybd., Farmington, New Mexico, June 12, 1965, and on *S. orcuttii* Rose, Picacho, Arizona, July 8, 1966, by D. M. T.

24

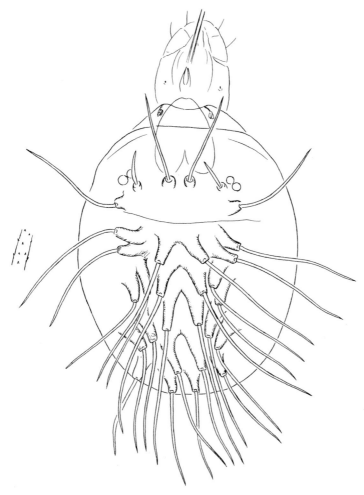

Figure 27, Dorsum of female.
Hystrichonychus ambiguae, new species

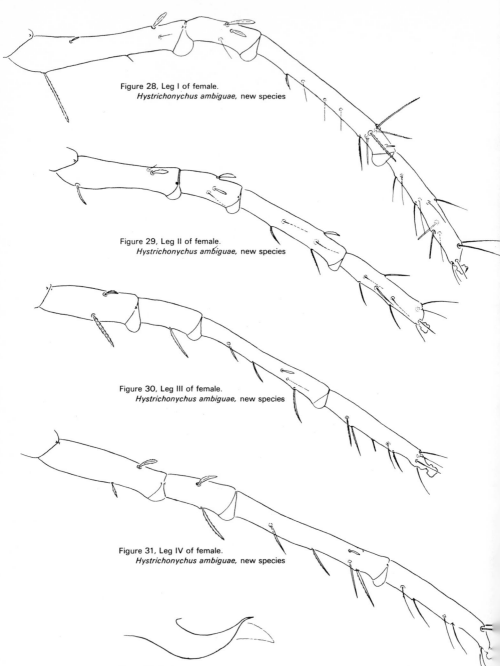

Figure 28, Leg I of female.
Hystrichonychus ambiguae, new species

Figure 29, Leg II of female.
Hystrichonychus ambiguae, new species

Figure 30, Leg III of female.
Hystrichonychus ambiguae, new species

Figure 31, Leg IV of female.
Hystrichonychus ambiguae, new species

Figure 32, Aedeagus.
Hystrichonychus ambiguae, new species

26

Hystrichonychus gracilipes (**Banks**)

(Figs. 33-37)

Tetranychus gracilipes Banks, 1900: 72.
Hystrichonychus gracilipes, Tuttle and Baker, 1964: 8.

The first pair of propodosomal setae of the female are almost on a transverse line with the second pair; the second pair of setae are much shorter than the other two pairs. The dorsal distal femoral I seta is strong and reaches to or surpasses the distal end of the segment. The serrations of the body setae are strong and obvious. The peritremes are branched distally. The male is similar to the female except the anterior pair of propodosomal setae are set far forward and in the normal position. The aedeagus is short, stout, upturned and narrowing abruptly posteriorly.

This species has been taken in Arizona on *Sphaeralcea* sp., *S. orcutti* Rose, *S. coccinea* (Pursh) Rydb., *Sida hederacea* (Dougl.) Torr., *Malva parviflora* L., and *Nolina microcarpa* Wats.

The figures given for *Hystrichonychus gracilipes* in Pritchard and Baker (1955) are in error and were drawn from specimens of *H. sidae*. *H. gracilipes* is here figured from material collected on *Sphaeralcea orcuttii* Rose, Dateland, Arizona.

Tetranycopsis Canestrini

Tetranycopsis Canestrini, 1889: 495, 504; Pritchard and Baker, 1955: 34; Wainstein, 1960: 116.

Type. Tetranychus horridus Canestrini and Fanzago.

The claws are long and padlike, each bearing two rows of short tenent hairs; the empodia resemble the claws. There are four pairs of propodosomal setae; the dorsal body setae are strong and set on prominent tubercles.

Members of this genus have not been found in Arizona.

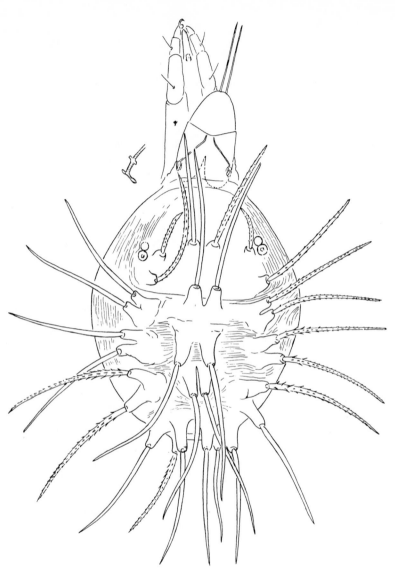

Figure 33, Dorsum of female.
Hystrichonychus gracilipes (Banks)

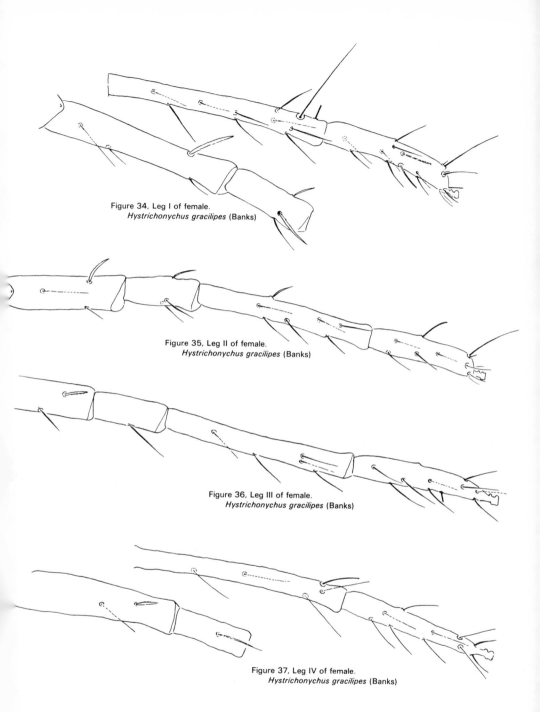

Figure 34, Leg I of female.
Hystrichonychus gracilipes (Banks)

Figure 35, Leg II of female.
Hystrichonychus gracilipes (Banks)

Figure 36, Leg III of female.
Hystrichonychus gracilipes (Banks)

Figure 37, Leg IV of female.
Hystrichonychus gracilipes (Banks)

Porcupinychus Anwarullah
(Fig. 38)

Porcupinychus Anwarullah, 1966: 71.
  Type. Porcupinychus abutilioni Anwarullah.

This mite possesses padlike claws and empodia. The peritreme is greatly enlarged distally. There are three pairs of propodosomal setae and eight pairs of hysterosomal setae, all setae being set on prominent tubercles. There are only four pairs of dorsocentral setae; the last two pairs are marginal. There are three pairs of marginal setae and one pair of humeral setae, which, in the female, is contiguous with the first pair of marginal setae. There are two sets of duplex setae on tarsus I and one set on tarsus II.

The small number of hysterosomal setae is distinctive.

The type is the only species known for the genus. It was collected in Pakistan.

Monoceronychus McGregor

Monoceronychus McGregor, 1945: 100; Pritchard and Baker, 1955: 74; Wainstein, 1960: 123.
  Type. Monoceronychus californicus McGregor.

The body is elongate and narrow. There are three pairs of propodosomal setae; the fourth pair of dorsocentral hysterosomal setae are widely separated and may even be marginal. There is a propodosomal and a hysterosomal shield; there are propodosomal projections over the rostrum. The claws and empodia are padlike.

KEY TO THE SPECIES OF MONOCERONYCHUS OF ARIZONA (FEMALES)

1. Fourth pair of dorsocentral setae marginal 2
   Fourth pair of dorsocentral setae not marginal 5
2. Posterior three pairs of hysterosomal setae slender 3
   Posterior three pairs of hysterosomal setae spatulate . . . . .
   . . . . *phleoides,* new species
3. Posterior three pairs of hysterosomal setae not similar to other marginal
   setae 4
   Posterior three pairs of setae similar to other marginal setae . . . .
   . . . . *scolus* Pritchard and Baker
4. Dorsal leg setae broad . . . . . *muhlenbergiae* Tuttle and Baker
   Dorsal leg setae slender . . . . *aristidae,* new species
5. Posterior three pairs of hysterosomal setae not similar to each other . 6
   Posterior three pairs of hysterosomal setae similar 7
6. Leg I longer than body . . . . . . *curtipendulae,* new species
   Leg I about as long as body . . . . *panici,* new species
7. All dorsal body setae broadly clavate 8
   All dorsal body setae not broadly clavate 9

30

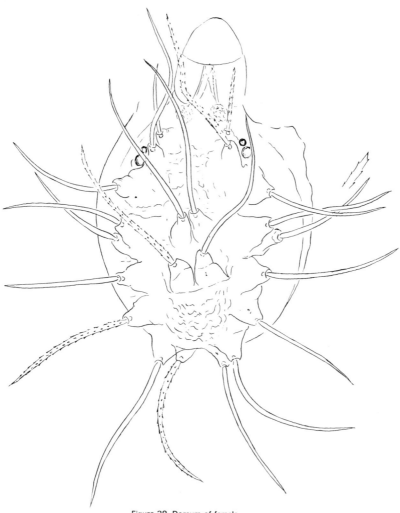

Figure 38, Dorsum of female.
Porcupinychus abutiloni Anwarullah

8. Hysterosoma with caudal three pairs of setae broadly . . . . . .
   spatulate . . . . *californicus* McGregor
   Hysterosoma with caudal three pairs of setae lanceolate . . . . . .
   . . . . *aechmetes* Pritchard and Baker
9. Anterior lateral projections of propodosoma much shorter than median
   projection . . . . *aristidoides,* new species
   Lateral projections equal in length to median projections . . . . . .
   . . . . *pulcherrimae* Tuttle and Baker

Monoceronychus phleoides, new species

(Figs. 39-43)

The posterior three pairs of hysterosomal setae are lanceolate to spatulate and are marginal; the other body setae are shorter and clavate. The anterior median propodosomal projection is much longer than the lateral projections.

Female. Rostrum short; stylophore abruptly narrowing at tip; peritreme ending in a simple bulb. Anterior median projections of propodosoma small but distinct, about as long as first pair of propodosomal setae, irregular in shape and rounded anteriorly; lateral projections almost non-existent, each bearing a clavate-serrate seta; other two pairs of propodosomal setae shorter and broader; dorsocentral portion of propodosoma with granulate pattern. Dorsal setae of hysterosoma, except for posterior three pairs, similar to the second and third pair of propodosomal setae; humeral setae smaller; posterior three pairs of hysterosomal setae lanceolate-serrate, about two and one-half times as long as others. Area of hysterosoma posterior to second pair of dorsocentral setae granulate. Leg I much shorter than body; dorsal setae of all legs short, strong, lanceolate-serrate; empodia about same length as true claws. Length of body 363 μ; including rostrum 421 μ.

Male. Not known.

Holotype. Female, *ex Lycurus phleoides* H.B.K. Portal, Arizona, August 28, 1964, by D. M. T.

A single nymph was also seen.

Monoceronychus scolus Pritchard and Baker

Monoceronychus scolus Pritchard and Baker, 1955: 90.

The three pairs of posterior hysterosomal setae are marginal; all marginal setae of the body are long, slender and set on tubercles; the first three pairs of dorsocentral setae are short and broadly lanceolate. The anterior median propodosomal projection is long and covers the rostrum.

Monoceronychus scolus has been taken in Arizona on *Aristida glabrata* (Vasey) Hitchc., *Bouteloua rothrockii* Vasey, *Hymenothrix wislizeni* Gray, *Pinus ponderosa* Lawson, and *Quercus gambelii* Nutt.

Monoceronychus muhlenbergiae Tuttle and Baker

Monoceronychus muhlenbergiae Tuttle and Baker, 1964: 18.

This species is similar to *Monoceronychus scolus,* but differs in having all but the posterior three pairs of hysterosomal setae short and broadly lanceolate; the posterior setae are also broader than those of *M. scolus.*

Monoceronychus muhlenbergiae has been collected in Arizona on *Muhlenbergia pulcherrima* Scribn.

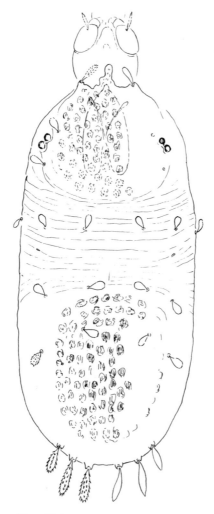

Figure 39, Dorsum of female.
Monoceronychus phleoides, new species

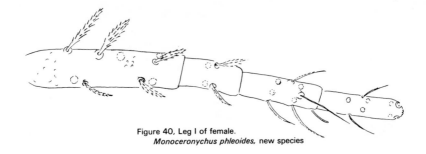

Figure 40, Leg I of female.
Monoceronychus phleoides, new species

Figure 41, Leg II of female.
Monoceronychus phleoides, new species

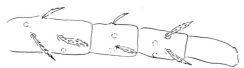

Figure 42, Leg III of female.
Monoceronychus phleoides, new species

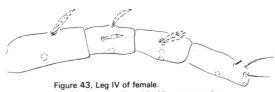

Figure 43, Leg IV of female.
Monoceronychus phleoides, new species

Monoceronychus aristidae, new species

(Figs. 44-45)

This species is distinctive in that the posterior three pairs of hysterosomal setae are marginal, slender, serrate, and set on tubercles. The dorsal leg setae are slender.

Female. Rostrum short, broad; stylophore narrowing anteriorly but not pointed; peritreme ending in a simple bulb. Anterior median propodosomal projection sharp, about four times as long as lateral projections, but not covering rostrum; lateral projections small, about as long as first pair of setae. Propodosomal setae broadly spatulate, finely serrate and subequal in length. Propodosoma with granulate pattern. All but posterior three pairs of hysterosomal setae similar to propodosomal setae; posterior pairs long, slender, and serrate. Hysterosoma with granulate pattern posterior to and between first pair of dorsocentral and lateral setae; area posterior to second pair of dorsocentral setae reticulate. Leg I shorter than body; dorsal leg setae short and narrowly lanceolate. Length of body 600 μ.

Male. Similar to female except posterior three pairs of hysterosomal setae more broadly lanceolate, and anterior median propodosomal projection rounded rather than pointed. Length of body 400 μ.

Holotype. Female, *ex Aristida adscensionis* L., Dateland, Arizona, May 19, 1964, by D. M. T.

Paratypes. Thirty females and two males with the above data.

35

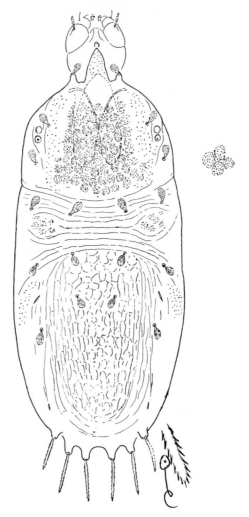

Figure 44, Dorsum of female.
Monoceronychus aristidae, new species

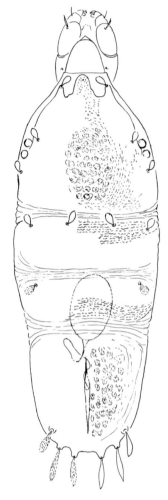

Figure 45, Dorsum of male.
Monoceronychus aristidae, new species

Monoceronychus curtipendulae, new species

(Figs. 46-50)

The fourth pair of dorsocentral hysterosomal setae are much more broadly clavate than the other two pairs of posterior setae. The first pair of legs are longer than the body.

Female. Rostrum short, broad; stylophore sharp distally; peritreme anastomosing distally. Anterior median propodosomal projection irregular, not reaching to end of first pair of propodosomal setae set on large tubercles. Anterior propodosomal setae longer than other two pairs, all broadly clavate; propodosomal integumentary pattern consists of large tubercles. Hysterosoma with tuberculate pattern posterior to second pair of dorsocentral setae. All dorsal hysterosomal setae, except posterior three pairs, similar to those on propodosoma. Fourth pair of dorsocentral setae broader than other two pairs of posterior setae; fifth pair of dorsocentral setae longer than others. Leg I longer than body. Dorsal leg setae narrowly lanceolate, serrate, as figured. Length of body 453 μ; including rostrum 530 μ.

Male. Not known.

Holotype. Female, *ex Bouteloua curtipendula* (Michx.) Torr., Globe, Arizona, June 27, 1962, by D. M. T.

Paratypes. Three females *ex Bouteloua aristidoides* (H.B.K.) Griseb., Tucson, Arizona, July 21, 1964, by D. M. T.

Other specimens have been collected in Arizona by D. M. T. on *Hymenothrix wislizeni* Gray, Marana, November 14, 1964; *Aristida glabrata* (Vasey) Hitchc., Oracle, July 21, 1964; *Bouteloua rothrockii* Vasey, Oracle, July 26, 1964; *Pinus ponderosa* Lawson, McNary, July 24, 1964; and *Quercus gambelii* Nutt. McNary, July 24, 1964.

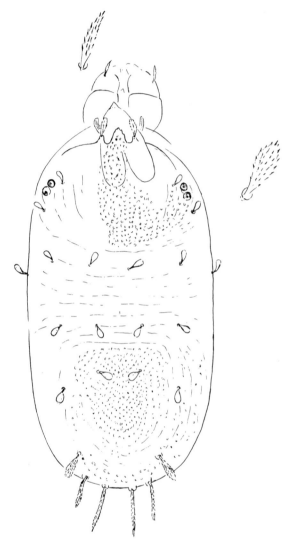

Figure 46, Dorsum of female.
Monoceronychus curtipendulae, new species

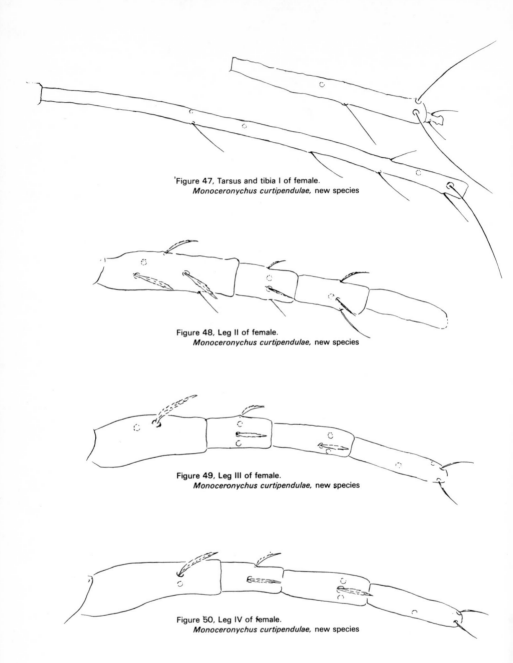

'Figure 47, Tarsus and tibia I of female.
Monoceronychus curtipendulae, new species

Figure 48, Leg II of female.
Monoceronychus curtipendulae, new species

Figure 49, Leg III of female.
Monoceronychus curtipendulae, new species

Figure 50, Leg IV of female.
Monoceronychus curtipendulae, new species

Monoceronychus panici, new species
(Figs. 51-54)

This species is distinctive in that the fourth pair of dorsocentral setae are not marginal, and are also shorter than the other two pairs of posterior setae. The propodosomal projections are subequal in length.

Female. Rostrum short, broad; stylophore short and broad and narrowing abruptly anteriorly; peritreme anastomosing distally. Anterior median propodosomal projection expanding distally and subequal in length to lateral projections. Anterior pair of propodosomal setae broadly lanceolate and subequal in length to other propodosomal setae. Propodosoma and posterior portion of hysterosoma with small tubercles. Hysterosomal setae, except for posterior two pairs, similar to those of propodosoma; posterior two pairs long, slender, and serrate. Dorsal leg setae of femur I lanceolate and slender. Leg I about as long as body. Length of body 414 μ; including rostrum 478 μ.

Male. Not known.

Holotype. Female, *ex Panicum obtusum* H.B.K., Portal, Arizona, August 28, 1964, by D. M. T.

Monoceronychus californicus McGregor

Monoceronychus californicus McGregor, 1945: 100; Tuttle and Baker, 1964: 16.

The fourth pair of dorsocentral setae are not marginal; all dorsal setae are similar, being broadly spatulate and serrate. The anterior median propodosomal projection is small, not covering the rostrum; the anterior pair of setae are set on large tubercles. The male is similar except in having a much smaller anterior median projection.

Monoceronychus californicus has been taken in Arizona on *Bouleloua curtipendula* (Michx.) Torr., *Distichlis stricta* (Torr.) Rydb., and *Muhlenbergia pulcherrima* Scribn.

Monoceronychus aechmetes Pritchard and Baker

Monoceronychus aechmetes Pritchard and Baker, 1955: 86.

This species is distinctive in that the anteromedian projection of the propodosoma is shorter than the rostrum, and in that the caudal three pairs of hysterosomal setae are broadly lanceolate.

It has been collected in Prescott, Arizona, on *Lycurus phleoides* H.B.K., August 18, 1965, by D. M. T. It was originally collected on various grasses: *Agrostis* sp., *Distichlis* sp., and *Cynodon dactylon* (L.) Pers. in California.

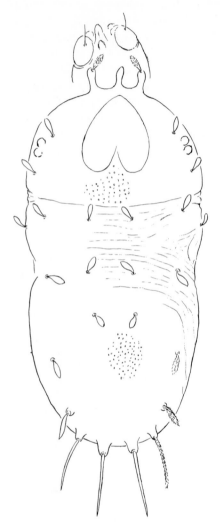

Figure 51, Dorsum of female.
Monoceronychus panici, new species

Figure 52, Propodosomal projection of female.
Monoceronychus panici, new species

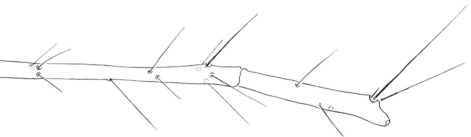

Figure 53, Tarsus and tibia I of female.
Monoceronychus panici, new species

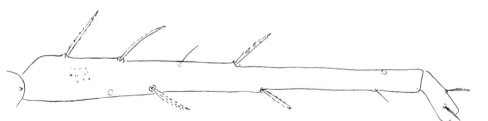

Figure 54, Femur I of female.
Monoceronychus panici, new species

Monoceronychus aristidoides, new species

(Figs. 55-59)

This species is distinctive in that the fourth pair of dorsocentral setae are not marginal and are similar to the other two posterior pairs. The dorsal body setae are not large nor broadly clavate. The lateral propodosomal projections are rudimentary.

Female. Rostrum short; stylophore sharp anteriorly; femoral seta of palpus short and broad distally; peritreme anastomosing distally. Propodosoma with tubercles; hysterosoma with tubercles posterior to second pair of dorsocentral setae. First pair of propodosomal setae lanceolate serrate, slightly larger than other propodosomal setae. Hysterosomal setae about the same size as propodosomal setae except for posterior two or three pairs which are about one-third larger. Dorsal leg setae short, lanceolate-serrate. Length of body 446 μ; including rostrum 480 μ.

Male. Not known.

Holotype. Female, *ex Bouteloua aristidoides* (H.B.K.) Griseb., Tucson, Arizona, July 21, 1964, by D. M. T.

Other specimens have been collected in Arizona by D. M. T. from *Chaenactis stevioides* Hook. and Arn., Gila Bend, May 1, 1964, and *Aristida adscensionis* L., Gila Bend, April 9, 1963.

Monoceronychus pulcherrimae Tuttle and Baker

Monoceronychus plucherrimae Tuttle and Baker, 1964: 18.

This species is distinctive in that the fourth pair of dorsocentral setae are not marginal; the posterior three pairs of hysterosmal setae are subequal in size and much larger than the other lanceolate body setae. The anterior median projection of the propodosoma is subequal in length to the lateral projections.

Monoceronychus pulcherrimae has been collected in Arizona on *Condalia lyciodes* (Gray) Weberb., *Muhlenbergia pulcherrima* Scribn., and *Pectis papposa* Harv. and Gray.

Reckia Wainstein

Reckia Wainstein, 1960: 154.

Type. Mesotetranychus samgoriensis Reck.

In this genus, the claws and empodia are padlike, the empodia being much longer than the claws bearing a double row of tenent hairs. The peritreme anastomoses distally. The fourth pair of dorsocentral hysterosomal setae are marginal; there are five pairs of dorsocentral setae and four pairs of lateral setae plus one pair of humeral setae; all setae are broadly clavate. The integument is covered by a tuberculate pattern rather than by striae.

Only one species, the type, is known. It was collected in the Transcaucasus.

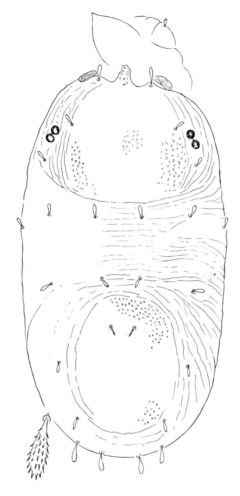

Figure 55, Dorsum of female.
Monoceronychus aristidoides, new species

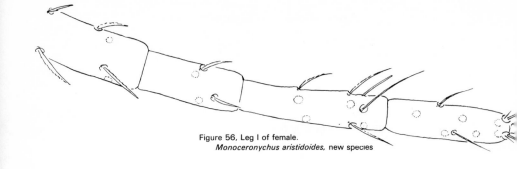

Figure 56, Leg I of female.
Monoceronychus aristidoides, new species

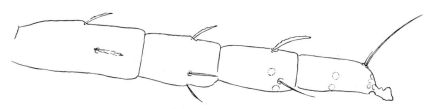

Figure 57, Leg II of female.
Monoceronychus aristidoides, new species

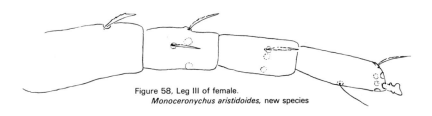

Figure 58, Leg III of female.
Monoceronychus aristidoides, new species

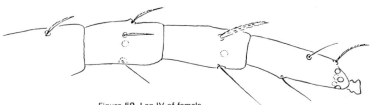

Figure 59, Leg IV of female.
Monoceronychus aristidoides, new species

Mesobryobia Wainstein

Mesobryobia Wainstein, 1956: 140.
Monoceronychus (*Mesobryobia*) Wainstein, 1960: 124.
  Type. Mesobryobia cervus Wainstein.

The claws and empodia are padlike and bear tenent hairs. The peritremes are elongate and anastomose distally. There are two projections over the rostrum, each bearing a seta; there is no medial projection. The fourth pair of dorsocentral setae are on strong tubercles and are further apart than the first three pairs, but are not quite marginal. The fourth laterals and fifth dorsocentrals are elongate, marginal and set on prominent tubercles.

Wainstein (1960) placed the following species in this genus: *cervus* Wainstein, *ter-poghossiani* Bagdasarian, and *corynetes* (Pritchard and Baker). The latter species is not typical, especially in lacking the strong posterior tubercles, and may belong elsewhere.

Members of this genus have not been found in Arizona.

Parapetrobia Meyer and Ryke

Parapetrobia Meyer and Ryke, 1959: 363.
  Type. Parapetrobia capensis Meyer and Ryke.

The claws and empodia are padlike. The peritreme is enlarged distally. There are three pairs of propodosomal setae and ten pairs of hysterosomal setae. Tarsus I possesses three pairs of duplex setae, and Tarsus II has the normal single set.

This genus, known only from the female, is South African. It has not been found in Arizona.

Aplonobia Womersley

Aplonobia Womersley, 1940: 252; Pritchard and Baker, 1955: 58;
  Wainstein, 1960: 139.
  Type. (*Aplonobia oxalis* Womersley) = *Aplonobia histricina* (Berlese).

The body setae are set on prominent tubercles and are well separated. The peritreme anastomoses distally. The legs of the female are long and slender and with normal number of setae. The body setae of the two known species are subequal in length.

There are two known species, *Aplonobia histricina* (Berlese) and *A. juliflorae,* new species. It is likely the latter can be expected in Arizona since the host plant *Prosopis juliflora* is very prevalent in southern areas of low elevation.

Aplonobia juliflorae, new species

(Figs. 60-61)

The dorsal body setae are set on small tubercles; the setae are of moderate length and subequal in length; the striae form a broken irregular pattern. The peritremes anastomose distally.

Female. Rostrum short; stylophore short, broad and rounded anteriorly; Peritremes branched distally. Prodosoma with few broken striae as figured; setae not longer than distance between bases, serrate and expanded distally; denticles of moderate strength. Hysterosomal setae similar to propodosomal setae, subequal in length except for shorter humeral setae; striae irregular as on propodosoma, transverse dorsally. Leg I as long as body; setation as figured, dorsal setae short, slightly lanceolate and serrate.

Male. Not known.

Holotype. Female, *ex Prosopis juliflora* (Swartz) DC., Alpine, California, July 26, by D. M. T.

Paratypes. Two females with the above data.

Although this species was taken in California, it is described here to represent the genus. The wide distribution of the host indicates that it eventually may be found in Arizona.

Paraplonobia Wainstein, new status

Aplonobia (*Paraplonobia*) Wainstein, 1960: 140.

Type. *Aplonobia* (*Paraplonobia*) *echinopsili* Wainstein.

The tarsal claws and empodia are padlike with tenent hairs. The dorsal body setae are not set on strong tubercles and are well separated. The setae may be short and broadly clavate, about as long as intervals between their bases and slightly clavate, or they may be slightly lanceolate; all setae are serrate. The propodosomal shield may be striated or tuberculate. The peritreme is simple in the type species, as well as in *Paraplonobia myops* (Pritchard and Baker), *P. tridens*, new species, and *P. hilariae*, new species. The peritreme anastomoses distally in *P. euphorbiae* (Tuttle and Baker), *P. prosopis* (Tuttle and Baker), *P. boutelouae*, new species, *P. dyschima* (Beer and Lang), and *P. corynetes* (Pritchard and Baker). Wainstein (1961) erected the genus *Langella* for *Aplonobia dyschima* Beer and Lang. We believe this to be only a subgenus of *Paraplonobia*.

KEY TO THE SPECIES OF PARAPLONOBIA IN ARIZONA (FEMALES)

1. With simple peritremes (*Paraplonobia* s. str.) 2
   Peritremes anastomosing distally (*Langella*) 4
2. Stylophore emarginate distally 3
   Stylophore rounded distally . . . . *myops* (Pritchard and Baker)

48

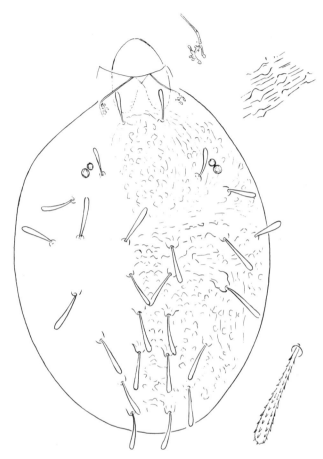

Figure 60, Dorsum of female.
Aplonobia juliflorae, new species

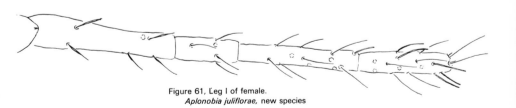

Figure 61, Leg I of female.
Aplonobia juliflorae, new species

49

3. Stylophore deeply emarginate; propodosomal shield with punctations .
. . . . *tridens,* new species
Stylophore slightly emarginate; propodosomal shield with striae . . .
. . . . *hilariae,* new species
4. Body setae strongly expanded distally 5
Body setae slender and lanceolate 6
5. All dorsal setae of tibia I subequal in length . *boutelouae,* new species
Dorsal distal seta of tibia I about two times as long as others . . . .
. . . . *prosopis* (Tuttle and Baker)
6. Propodosomal shield with striae forming a basket weave pattern . . .
. . . . *euphorbiae* (Tuttle and Baker)
Propodosomal shield with normal striae . *coldeniae* (Tuttle and Baker)

Paraplonobia (*Paraplonobia*) Wainstein

Type. Aplonobia (*Paraplonobia*) *echinopsili* Wainstein.

This subgenus is distinctive in having simple peritremes.

Paraplonobia (*Paraplonobia*) *myops* (Pritchard and Baker), new combination

Aplonobia myops Pritchard and Baker, 1955: 63; Tuttle and Baker, 1964: 11.
Paraplonobia myops, Wainstein, 1960: 141.

This species is distinctive in having slender, serrate setae which are not as long as the distance between their bases; the setae are not set upon tubercles. The first pair of legs are about as long as the body; the empodial pads are much longer than the pads of the claws; the stylophore is rounded anteriorly, and the peritreme ends in a simple bulb.

Paraplonobia myops has been collected in Arizona on *Ambrosia dumosa* (Gray) Payne, *Distichlis stricta* (Torr.) Rydb., *Heterotheca subaxillaris* (Lam.) Britt. and Rusby, *Hilaria rigida* (Thurb.) Benth., and *Tridens pulchellus* (H.B.K.) Hitchc.

Paraplonobia (*Paraplonobia*) *tridens,* new species
(Figs. 62-66)

This species is distinctive in having short, slightly lanceolate-serrate setae which are not set on tubercles, in having a simple peritreme, and in having a deeply cleft stylophore.

Female. Rostrum short and broad; stylophore long, broad and incised anteriorly; peritreme simple distally. All dorsal body setae short, not as long as distances between bases, slightly lanceolate and serrate, not set on tubercles and all subequal in length. Propodosoma with small punctations; hysterosoma with transverse striae. Leg I slightly longer than body; all leg setae of medium length and slender; tarsi III and IV each with a slender distal solenidion; legs I and II as figured. Length of body 413 μ; including rostrum 467 μ.

Male. Stylophore cleft anteriorly; peritreme simple distally. Setae similar to those of female; aedeagus typical for genus. Tarsus I with six sets of duplex setae; tibia I with four sets of duplex setae; Tarsus II normal, with one set of duplex setae; tibia II without duplex setae; tarsi III and IV each with a distal solenidion. Length of body 370 μ.

Holotype. Female. *ex Tridens pulchellus* (H.B.K.) Hitchc., Portal, Arizona, August 28, 1964, by D. M. T.

Paratypes. Nine females and two males with the above data.

Specimens were also taken on the above host by D. M. T. at Gila Bend, Arizona, October 15, 1963, and *Celtis reticulata* Torr., Portal, Arizona, August 28, 1964.

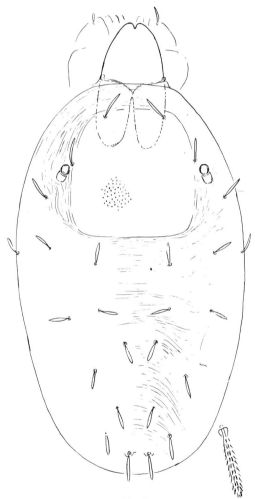

Figure 62, Dorsum of female.
Paraplonobia (Paraplonobia) tridens, new species

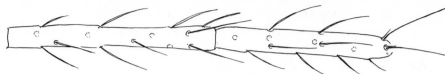

Figure 63, Tarsus and tibia I of female.
Paraplonobia (Paraplonobia) tridens, new species

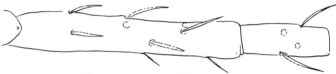

Figure 64, Genu and femur I of female.
Paraplonobia (Paraplonobia) tridens, new species

Figure 65, Leg II of female.
Paraplonobia (Paraplonobia) tridens, new species

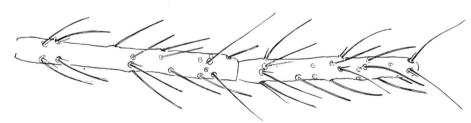

Figure 66, Tarsus and tibia I of male.
Paraplonobia (Paraplonobia) tridens, new species

Paraplonobia (*Paraplonobia*) *hilariae,* new species
(Figs. 67-70)

This species has short serrate dorsal body setae which are not set on tubercles, and the striae of the propodosoma are longitudinal.

Female. Rostrum more elongate than in other species of the genus; stylophore lightly indentate anteriorly; peritreme simple distally. Propodosoma

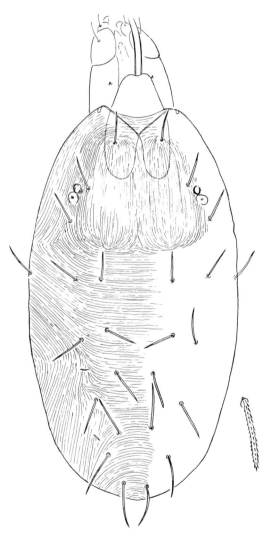

Figure 67, Dorsum of female.
Paraplonobia (Paraplonobia) hilariae, new species

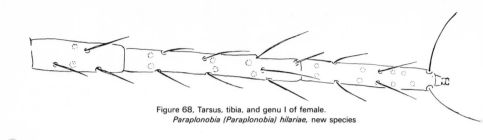

Figure 68, Tarsus, tibia, and genu I of female.
Paraplonobia (Paraplonobia) hilariae, new species

Figure 69, Femur I of female.
Paraplonobia (Paraplonobia) hilariae, new species

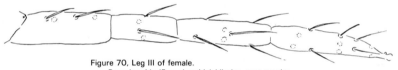

Figure 70, Leg III of female.
Paraplonobia (Paraplonobia) hilariae, new species

with longitudinal striae; hysterosoma with transverse striae. All dorsal setae except posterior hysterosomal setae shorter than distances between their bases, not set on tubercles, slender and serrate. Leg I as long as body; dorsal setae short and slender; tarsi III and IV each with a distal slender solenidion. Length of body 466 μ; including rostrum 530 μ.

Male. Not known.

Holotype. Female, *ex Hilaria rigida* (Thurb.) Benth., Gila Bend, Arizona, October 15, 1963, by D. M. T.

Paratypes. Twelve females with the above data.

Paraplonobia (*Langella*) Wainstein, new status

Langella Wainstein, 1961: 607.

Type. *Aplonobia dyschima* Beer and Lang.

This subgenus is distinctive in possessing anastomosing peritremes.

Paraplonobia (*Langella*) *boutelouae*, new species
(Figs. 71-72)

This species is distinctive in lacking setal tubercles; in having the peritremes anastomosing distally; in having irregular broken striae on the propodosoma and irregular striae on the hysterosoma; in that the setae are broadly clavate; and in that all dorsal setae or tibia I are subequal in length.

Female. Rostrum short; stylophore short, broad and rounded anteriorly; peritremes anastomosing distally. Propodosoma with broken longitudinal

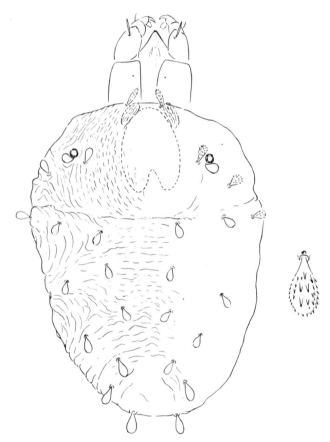

Figure 71, Dorsum of female.
Paraplonobia (Langella) boutelouae, new species

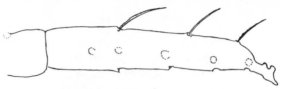

Figure 72, Tarsus IV of female.
Paraplonobia (Langella) boutelouae, new species

striae; hysterosoma with irregular transverse striae. All dorsal body setae subequal in length, strongly spatulate and serrate. Empodia at least two to three times as long as pads of true claws. Leg I not quite as long as body; all dorsal leg setae short; spatulate and subequal in length. Length of body 333 μ; including rostrum 413 μ.

Male. Not known.

Holotype. Female, *ex Bouteloua curtipendula* (Michx.) Torr., Portal, Arizona, August 28, 1964, by D. M. T.

Paraplonobia (Langella) prosopis (Tuttle and Baker), new combination

Aplonobia prosopis Tuttle and Baker, 1964: 13.

Paraplonobia prosopis is similar to *P. boutelouae,* new species. It differs, however, in having longer dorsal body setae, and in that the anterior dorsal setae of tibia I is almost twice as long as the other dorsal setae. The rostrum is short, the stylophore is short and rounded anteriorly, and the peritreme is anastomosing distally. The striae are longitudinal, broken on the propodosoma, and transverse on the hysterosoma.

This species was taken in Arizona on *Prosopis juliflora* (Swartz) DC.

Paraplonobia (Langella) euphorbiae (Tuttle and Baker), new combination

Aplonobia euphorbiae Tuttle and Baker, 1964: 11.

This species is distinctive in that the dorsal body setae are subequal in length, not longer than the distances between their bases; the setae are not set on tubercles. The striae form a basketweave pattern on the propodosoma, and an irregular transverse pattern on the hysterosoma. The peritremes anastomose distally.

Paraplonobia euphorbiae has been collected in Arizona on *Euphorbia albomarginata* Torr. and Gray and *Polygonum argyrocoleon* Steud.

Paraplonobia (*Langella*) *coldeniae* (**Tuttle and Baker**), **new combination**

Aplonobia coldeniae Tuttle and Baker, 1964: 16.

Paraplonobia coldeniae is distinctive in having an extremely elongate rostrum; in that all the dorsal body setae taper distally and are longer than the distances between their bases; in having the peritreme anastomosing distally; and in lacking setal tubercles.

This species has been taken in Arizona on *Coldenia palmeri* Gray.

Anaplonobia Wainstein

Aplonobia (*Anaplonobia*) Wainstein, 1960: 143.

Type. Aplonobia calame Pritchard and Baker.

This genus is distinctive in that the body setae are not set on prominent tubercles and all setae are well separated. The fourth pair of dorsocentral hysterosomal setae, although separated, are closer together than the first three pairs of dorocentral setae. The peritremes anastomose distally. Both males and females have legs shorter than the bodies. There is a normal complement of duplex setae on tarsi I and II, although the male possesses many solenidia on tarsus I and tibia I. The leg setae have few, but long, branches.

Two species are known, *Anaplonobia calame* (Pritchard and Baker), and *A. acharis* (Pritchard and Baker). Both were described from California but have not been found in Arizona.

Neopetrobia Wainstein

Neopetrobia Wainstein, 1956: 151.

Type. Neopetrobia dubinini Wainstein,

This genus is closely related to *Anaplonobia*. The tarsal claws and empodia are padlike, bearing tenent hairs. The peritreme is elongate and anastomosing distally. The body setae are short and slender. The fourth pair of dorsocentral hysterosomal setae are set further apart than the first three pairs, but are not marginal. The tactile leg setae are slender and without obvious spination.

Wainstein lists *Neopetrobia dubinini* Wainstein, *N. mcgregori* (Pritchard and Baker), and *N. vediensis* (Bagdasarian) as belonging to this genus. These are not known to occur in Arizona.

Beerella Wainstein

Beerella Wainstein, 1961: 606.

   Type. Aplonobia verrucosa Beer and Lang.

   This genus is distinctive in that the dorsal setae are strong and set on strong tubercles; there are only four pairs of dorsocentral setae, all pairs of which are contiguous. The peritremes anastomose distally. The leg setation is normal; the number of duplex setae in the female is normal; the male is not known. The claws and empodia are padlike.

   This genus is known only from the type which was collected in Oaxaca, Mexico, on *Lippia graveolens* H. B. K.

Georgiobia Wainstein

Georgiobia Wainstein, 1960: 138.

   Type. Petrobia shirakensis Reck.

   This genus is distinctive in that the humeral setae are set on strong tubercles and are anterior to or contiguous with the first lateral hysterosomal setae. There are thirteen pairs of hysterosomal setae, all on strong tubercles; the second, third, and fourth pair of dorsocentral setae are contiguous; the lateral setae are marginal or nearly so. The empodium is padlike, elongate and with a row of tenent hairs; the claws are padlike and each claw possesses a single pair of tenent hairs. The peritreme is anastomosing distally. The setae of the male are not set on strong tubercles, and are well separated.

   This genus was described from southern Russia. Several species have been found in Arizona.

KEY TO THE SPECIES OF GEORGIOBIA IN ARIZONA (FEMALES)

1. Dorsocentral setae one to three not contiguous nor on strong tubercles   2
   Dorsocentral setae two to four contiguous; all dorsocentral setae on strong tubercles   5
2. Dorsocentral setae one to three small and subequal in length   3
   Dorsocentral setae one and four much longer than two and three . .   4
3. Dorsocentral setae one to three minute, not nearly as strong as other dorsal body setae . . . . *ambrosiae,* new species
   Dorsocentral setae one to three at least half as long as distance between their bases, strong, half as long or equal to length of other dorsal body setae . . . . *haplopappi,* new species
4. Dorsal body setae broadly spatulate . . . . *dyssodiae,* new species
   Dorsal body setae slender . . . . . *anisa* (Pritchard and Baker)
5. Dorsocentral setae short, reaching about halfway to margin of body .   6
   Dorsocentral setae long, reaching margin of body . . . .
   . . . . *deina* (Pritchard and Baker)
6. Leg I as long as body . . . . *potentillae,* new species
   Leg I shorter than body . . . . *sphaeralceae,* new species

Georgiobia ambrosiae, new species

(Figs. 73-76)

This species is distinctive in that the first three pairs of dorsocentral hysterosomal setae are minute, as are many of the leg setae.

Male. Dorsal setae of genu and femur of palpus short and lanceolate serrate; stylophore short, broad and rounded anteriorly; peritremes anastomosing distally. Propodosomal setae of medium length, serrate, stout and

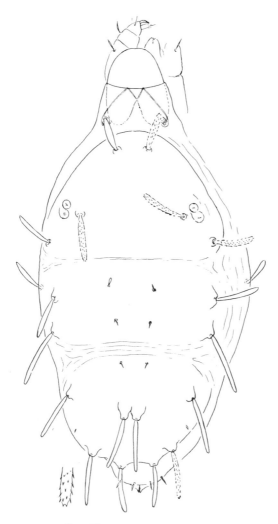

Figure 73, Dorsum of male.
Georgiobia ambrosiae, new species

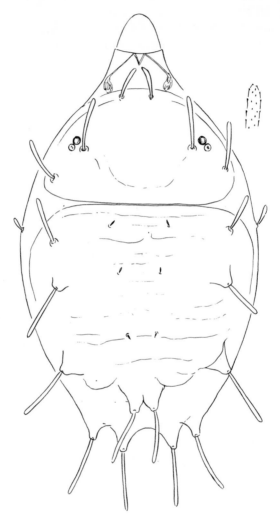

Figure 74, Dorsum of female.
Georgiobia ambrosiae, new species

rounded distally. Marginal and fourth pair of dorsocentral hysterosomal setae longer and more slender than propodosomal setae; first three pairs of dorso-central setae minute, the first pair varying in size, and usually slightly larger than other two pairs; fourth pair of dorsocentral setae long, strong and set on prominent contiguous tubercles; marginal setae all long and subequal in length to fourth pair of dorsocentral setae. Aedeagus straight, long, slender and abruptly narrowing distally. Legs are distinctive in that the dorsal setae may be minute, as figured. Length of body 300 μ; including rostrum 395 μ.

Female. Similar to male. Length of body 427 μ; including rostrum 542 μ.

Holotype. Male, *ex Ambrosia confertiflora,* Prescott, Arizona, August 19, 1965, by D. M. T.

Paratypes. Eleven females, two males and numerous immatures with the above data.

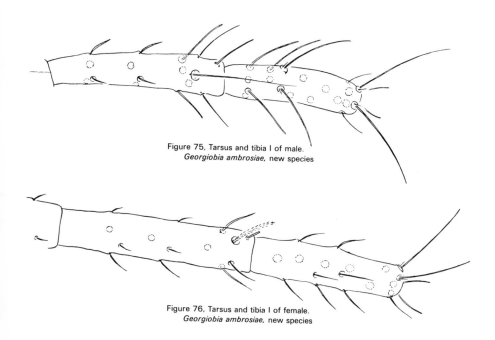

Figure 75, Tarsus and tibia I of male.
Georgiobia ambrosiae, new species

Figure 76, Tarsus and tibia I of female.
Georgiobia ambrosiae, new species

Georgiobia haplopappi, new species

(Figs. 77-78)

Georgiobia haplopappi is distinctive in having the posterior four pairs of hysterosomal setae longer than the others; the dorsocentral setae one to three are of equal length but shorter than the lateral setae.

Female. Rostrum short; stylophore short, broad and rounded anteriorly; peritreme ending in an elongate anastomosing process. Propodosoma and hysterosoma with few irregular and faint striae. Propodosomal setae clublike, serrate, the third pair slightly smaller than others. First transverse row of hysterosomal setae short and subequal in length; humeral setae shorter; first three pairs of dorsocentral setae subequal in length and shorter than posterior setae; second pair of lateral setae longer than dorsocentrals but shorter than posterior setae; posterior four pairs of hysterosomal setae slightly clublike, serrate and longer than other setae. Denticles of setae minute. Leg I as long as body; dorsal leg setae short, slender and serrate; empodial pad longer than pads of true claws. Length of body 426 μ; including rostrum 440 μ.

Male. Not known.

Holotype. Female, *ex Haploppapus gracilis* Nutt., McNary, Arizona, August 14, 1963, by D. M. T.

A single nymph with the above data is in the collection.

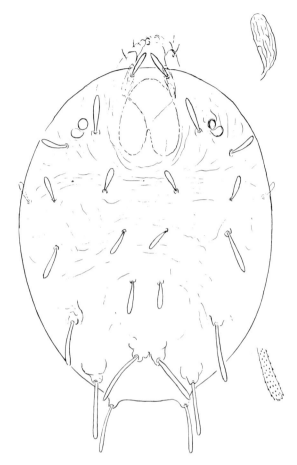

Figure 77, Dorsum of female.
Georgiobia haplopappi, new species

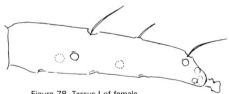

Figure 78, Tarsus I of female.
Georgiobia haplopappi, new species

Georgiobia dyssodiae, new species

(Figs. 79-81)

The dorsal setal pattern of *Georgiobia dyssodiae* is similar to that of *G. anisa* (Pritchard and Baker), but differs in that the posterior four pairs of hysterosomal setae are strong and expanded distally.

Female. Rostrum short; stylophore short, broad and rounded anteriorly; peritremes anastomosing distally; the enlargement elongate. Propodosoma

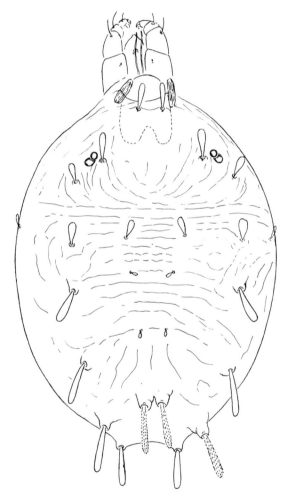

Figure 79, Dorsum of female.
Georgiobia dyssodiae, new species

64

with few striae, propodosomal setae subequal in length, expanded distally, with few small serrations. Humeral setae of hysterosoma very short but expanded distally; first pair of dorsocentral setae slightly smaller than first pair of lateral setae; second and third pair of dorsocentral setae minute, serrate and clavate. Other dorsal setae long, strong and broadened distally. Leg I about as long as body; segments of leg I relatively longer than in *Georgiobia anisa;* empodium two times as long as true claws; true claws with a pair of tenent hairs and empodium with a double row of tenent hairs. Length of body 480 μ; including rostrum 546 μ.

Male. Not known.

Holotype. Female, *ex Dyssodia papposa* (Vent.) Hitchc., Pueblo, Colorado, September 6, 1963, by D. M. T.

Although this species was collected in Colorado, the host plant is found in Arizona, and, consequently, we have included this species here.

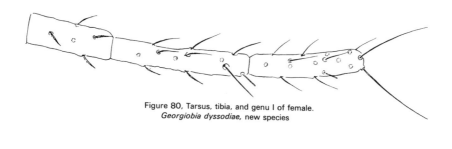

Figure 80, Tarsus, tibia, and genu I of female.
Georgiobia dyssodiae, new species

Figure 81, Femur I of female.
Georgiobia dyssodiae, new species

Georgiobia anisa (Pritchard and Baker), new combination

Aplonobia anisa Pritchard and Baker, 1955: 64; Wainstein, 1960: 140; Tuttle and Baker, 1964: 11.

This species is distinctive in having the second and third pairs of dorsocentral setae small and lanceolate-serrate, in that the posterior four pairs of hysterosomal setae are long, slender and serrate, and in that the other body setae are shorter, but strongly lanceolate serrate. The female has the normal number of duplex setae. Tarsus I of the male possesses four sets of duplex setae.

Georgiobia anisa has been collected in Arizona on *Ambrosia confertiflora* DC., *Atriplex confertifolia* (Torr. and Frém.) Wats., *Chrysopsis villosa* (Pursh) Nutt., *Dicoria canescens* Gray, *Fraxinus velutina* Torr., *Haplopappus gracilus* Nutt., *Parthenocissus quinquefolia* (L.) Planch., and *Viguiera multiflora* (Nutt.) Blake.

Georgiobia deina (Pritchard and Baker), new combination

Aplonobia deina Pritchard and Baker, 1955: 67; Wainstein, 1960: 140; Tuttle and Baker, 1964: 11.

This species is distinctive in having long setae set on strong tubercles, the tubercles of the second, third and fourth pair of dorsocentral setae being contiguous. The peritremes anastomose distally. The empodium is about as long as the pad of the true claw.

Georgiobia deina has been collected in Arizona on *Ambrosia deltoidea* (Torr.) Payne and *Verbesina encelioides* (Cav.) Benth. and Hook.

66

Georgiobia potentillae, new species

(Figs. 82-84)

This species is related to *Georgiobia sphaeralceae,* new species, but differs in having much thinner dorsal body and leg setae, in having stronger denticles on these setae, and in that leg I is as long as the body.

Female. Rostrum short; stylophore short, broad and rounded anteriorly; peritremes anastomosing distally. Propodosoma with few irregular striae;

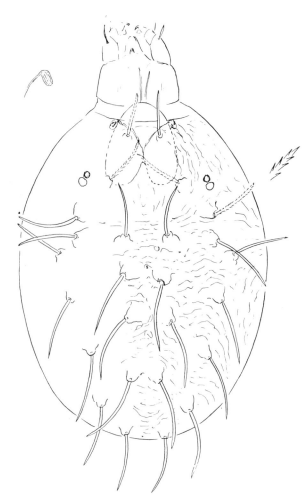

Figure 82, Dorsum of female.
Georgiobia potentillae, new species

all three pairs of setae subequal in length, serrate and slender. Hysterosomal and humeral setae slightly longer than propodosomal setae, subequal in length, slender and serrate. Hysterosomal setae on strong tubercles; each pair of dorsocentral tubercles two to four contiguous; striae few, irregular and transverse. Leg I as long as body; leg setae serrate and slender and not as strong as in *Georgiobia sphaeralceae*. Length of body 530 μ; including rostrum 640 μ.

 Male. Not known.

 Holotype. Female, *ex Potentilla hippiana* Lehm., McNary, Arizona, June 29, 1962, by D. M. T.

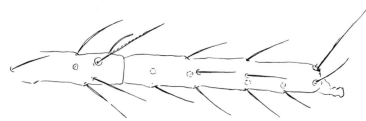

Figure 83, Tarsus and tibia I of female.
Georgiobia potentillae, new species

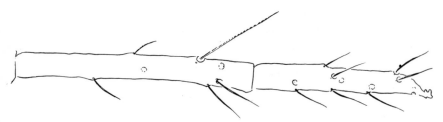

Figure 84, Tarsus and tibia IV of female.
Georgiobia potentillae, new species

Georgiobia sphaeralceae, new species

(Figs. 85-87)

This species is distinctive in having medium length setae set on strong tubercles, the second to fourth pair of dorsocentral setae being contiguous; the denticles of the setae are numerous and small. The peritremes anastomose distally. Legs I are not as long as the body.

Female. Rostrum short; stylophore broadly rounded anteriorly; peritremes anastomosing distally, the enlargements round. Propodosoma with few irregular striae; anterior pair of propodosoma setae shorter than others

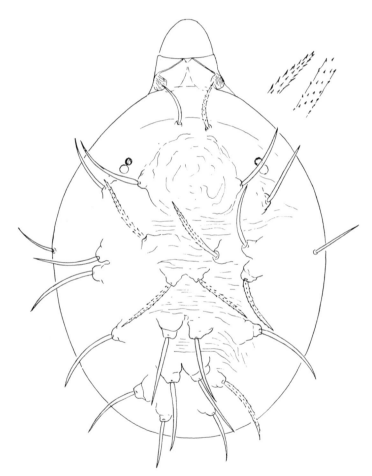

Figure 85, Dorsum of female.
Georgiobia sphaeralceae, new species

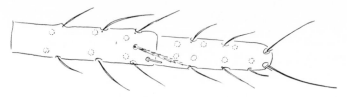

Figure 86, Tarsus and tibia I of female.
Georgiobia sphaeralceae, new species

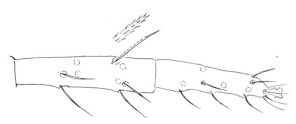

Figure 87, Tarsus and tibia IV of female.
Georgiobia sphaeralceae, new species

which are subequal in length; all dorsal setae strong, slightly lanceolate. Humeral setae smaller than others; hysterosomal setae similar to those of propodosoma except for shorter fifth pair of dorsocentral setae. First pair of dorsocentral setae well separated; second, third and fourth pairs contiguous; hysterosomal tubercles very strong; striae few, weak and transverse. Leg I not as long as body. Leg setae of medium length, slightly lanceolate-serrate. Length of body 480 μ; including rostrum 533 μ.

Male. Not known.

Holotype. Female, *ex Sphaeralcea ambigua* Gray, Camp Verde, Arizona, February 28, 1963, by D. M. T.

Paratypes. One female with the above data; two females *ex Ambrosia deltoidea* (Torr.) Payne with the same collecting data as above.

Schizonobiella Beer and Lang

Schizonobiella Beer and Lang, 1958: Wainstein, 1960: 139.

Type. Schizonobiella aeola Beer and Lang.

The true claws are padlike, each bearing a single pair of tenent hairs; the empodium is strong and hooked, bearing on each side a row of hairs anastomosing distally to form a terminal tenent. There is only a single set of duplex setae on tarsus I; tibia I of the male also bears a single set of duplex setae. The peritremes anastomose distally.

This species has been found on grass near Alice, Texas.

PETROBIINI RECK

Petrobiinae Reck, 1952: 423.
Petrobiini, Pritchard and Baker, 1955: 42; Wainstein, 1960: 131.
   This tribe is characterized by having padlike claws and uncinate empodia.
The following three genera belong here:
   Petrobia Murray
   Schizonobia Womersley
   Mezranobia Athias-Henriot

Petrobia Murray

Petrobia Murray, 1877: 118; Wainstein, 1960: 133.
   Type. (*Trombidium lapidum Hammer*) = *Petrobia latens* (Müller).

   There are three pairs of propodosomal setae, nine pairs of hysterosomal
setae and a pair of humeral setae. The true claws are padlike with tenent
hairs; the empodium is uncinate and has two rows of ventrally directed
tenent hairs.
   Wainstein (1960) subdivided *Petrobia* into the following subgenera:
Petrobia Murray
Tentranychina Banks
Mesotetranychus Reck
Each of these is represented in Arizona by a single species.

KEY TO THE SUBGENERA AND SPECIES OF PETROBIA IN ARIZONA
1. Dorsal setae not set on tubercles   2
    Dorsal setae, or some dorsal setae, set on large tubercles; peritreme
    hooked distally *(Tetranychina)* . . . . *harti* (Ewing)
2. Peritreme anastomosing distally *(Petrobia)* . . . *latens* (Müller)
    Peritreme hooked distally *(Mesotetranychus)* . . . .
    . . . .*phaceliae* (Tuttle and Baker)

Petrobia (Petrobia) Murray

Petrobia Murray, 1877: 118; Wainstein, 1960: 133.
   Type. (*Trombidium lapidum Hammer*) = *Acarus latens* Müller.

Petrobia (Petrobia) latens Muller

Acarus latens Müller, 1776: 187.
Petrobia latens, Pritchard and Baker, 1955: 51; Wainstein, 1960: 134;
   Tuttle and Baker, 1964: 9.
   The dorsal setae of the female (there are no males) are shorter than
the distances between their bases, and are not set on tubercles; the anterior
pair of propodosomal setae are the longest. The first pair of legs are longer

71

than the body. The peritremal enlargement is slender and longer than broad.

Petrobia latens is found on many plants throughout various parts of the world. At times they invade homes, becoming a nuisance. Arizona records include: *Abronia villosa* Wats., *Aegilops cylindrica* Host, *Allium cepa* L., *Astragalus arizonicus* Gray, *Bebbia juncea* (Benth.) Greene, *Bromus arizonicus* (Shear) Stebbins, *Cenchrus echinatus* L., *Cichorium endivia* L., *Coldenia palmeri* Gray, *Convolvulus arvensis* L., *Cucumis melo* L., *Cynodon dactylon* (L.) Pers., *Dalea albiflora* Gray, *Daucus carota* L., *Echinochloa crusgalli* (L.) Beauv., *Eschscholtzia mexicana* Greene, *Ficus carica* L., *Gladiolus hortulanus* Bailey, *Gossypium hirsutum* L., *Haploppapas spinulosus* (Pursh) DC., *Hordeum vulgare* L., *Iris missouriensis* Nutt., *Lactuca sativa* L., *Lepidium thurberi* Wooton, *Lesquerella gordoni* (Gray) Wats., *Lupinus* sp., *Medicago sativa* L., *Oenothera clavaeformis* Torr. and Frém., *Oenothera primiveris* Gray, *Phalaris minor* Ratz., *Polygonum aviculare* L., *Potentilla norvegica* L., *Sitanion hystrix* (Nutt.) J. G. Smith, *Tragopogon porrifolius* (L.), and *Triticum aestivum* L.

Petrobia (*Tetranychina*) Banks

Tetranychina Banks, 1917; Pritchard and Baker, 1955: 43.
Petrobia (*Tetranychina*), Wainstein, 1960: 136.
  Type. Terranychina apicalis Banks.

Petrobia (*Tetranychina*) *harti* (Ewing)

Neophyllobius harti Ewing, 1909: 405.
Petrobia harti, Pritchard and Baker, 1955: 45; Wainstein, 1960: 137;
  Tuttle and Baker, 1964: 8.

The dorsal setae of the female are long and set on strong tubercles; the fifth pair of dorsocentral setae are much shorter than the others. The first pair of dorsocentral setae of the male are comparatively long and slender, and the other dorsocentral setae are short and lanceolate. The first pair of legs of the female are about twice as long as the body; the first pair of legs of the male are about three times as long as the body.

Petrobia harti has been taken on *Oxalis* throughout various parts of the world and has been taken on that host in Arizona.

Petrobia (*Mesotetranychus*) Reck

Mesotetranychus Reck, 1948: 178.
Petrobia (*Mesotetranychus*) Wainstein, 1960: 137.
  Type. Mesotetranychus vachustii Reck.

72

Petrobia (*Mesotetranychus*) *phaceliae* (Tuttle and Baker), new combination

Petrobia phaceliae Tuttle and Baker, 1964: 9.

This species has the long free anastomosing peritremes of *Petrobia latens*. The dorsal body setae, however, are longer than the distance between their bases and are not set upon tubercles.

This mite was collected on *Phacelia* sp. in Arizona.

Schizonobia Womersley

Schizonobia Womersley, 1940: 251; Pritchard and Baker, 1955: 56; Wainstein, 1960: 138.

Type. Schizonobia sycophanta Womersley.

The true claws are padlike, the empodium is uncinate, all bearing a single pair of tenent hairs. In the only known species, the peritreme is broadly expanded distally.

This genus is represented by a single species from Tasmania (Australia).

Mezranobia Athias-Henriot

Mezranobia Athias-Henriot, 1961: 2.

Type. Mezranobia vannatum Athias-Henriot.

The true claws are padlike; the empodia are uncinate; both claws and empodia bear tenent hairs. The peritreme is expanded and free distally. There are three propodosomal projections over the rostrum, each bearing a seta.

This genus, based upon a single species, is known only from Algiers.

NEOTRICHOBIINI, NEW TRIBE

Neotrichobiini is similar to the Petrobiini in having a clawlike empodium and padlike claws. It is distinctive in having in all stages, except the larva, many ventral body and coxal setae. There are seven shields on the dorsum of the adult. The included genus is:

Neotrichobia, new genus.

Neotrichobia, new genus

Type. Neotrichobia arizonensis, new species.

There are three pairs of propodosomal setae and nine pairs of hysterosomal setae which are arranged on seven dorsal shields. The venter of the adults and nymphs is covered with many setae; the female has three pairs of genital and three pairs of anal setae. The male has five pairs of genito-anal setae. The larva has only three pairs of ventral setae, and only coxal I possesses a seta.

The only species for the genus, *N. arizonensis*, new species, is found in Arizona.

Neotrichobia arizonensis, new species
(Figs. 88-93)

The numerous ventral setae are distinctive.

Female. Propodosoma with three pairs of setae; hysterosoma with nine pairs of setae; with one large median and two small lateral shields. Hysterosoma with two large median and two smaller lateral shields. Peritremes anastomosing distally. Palpal femoral and genual setae plumose. Ventral body setae plumose; instead of the usual three pairs of ventral setae about 38 pairs of plumose setae present; each coxa with from eight to fifteen plumose setae. With three pairs of simple genital setae and three pairs of simple anal setae; with two pairs of plumose para-anal setae. Legs with many plumose setae; tarsus I with three short slender solenidia and a pair of duplex setae; tibia I with two solenidia. Tarsus II without duplex setae, but tarsus II and tibia II each with a single solenidion. Tarsus III and IV each with two solenidia. Length of body 466 μ; including rostrum 600 μ.

Male. Similar in setation pattern to female. With five pairs of genito-anal setae; aedeagus long, strong and slightly curved. Length of body 386 μ; including rostrum 480 μ.

Nymph. Setal pattern similar to female, but without dorsal shields.

Larva. Normal for the Bryobiinae in having two pairs of ventral setae and a single seta on coxa I; all setae plumose.

Holotype. Female, *ex Aristida adscensionis* L., Tucson, Arizona, June 24, 1964, by D. M. T.

Paratypes. Male, nymph, and larva with the above data.

74

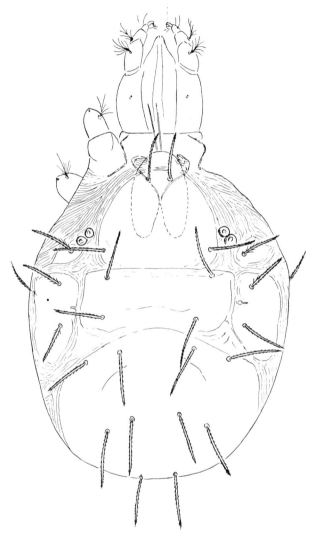

Figure 88, Dorsum of female.
Neotrichobia arizonensis, new species

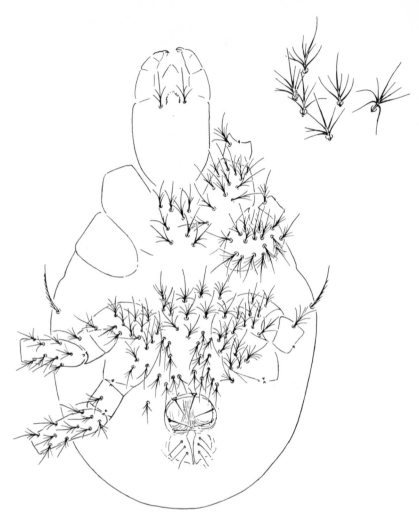

Figure 89, Venter of female.
Neotrichobia arizonensis, new species

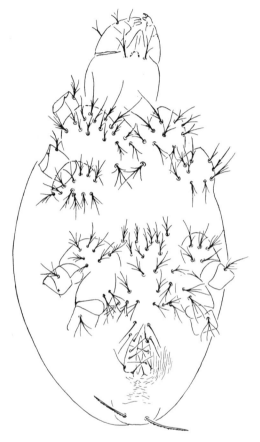

Figure 90, Venter of nymph.
Neotrichobia arizonensis, new species

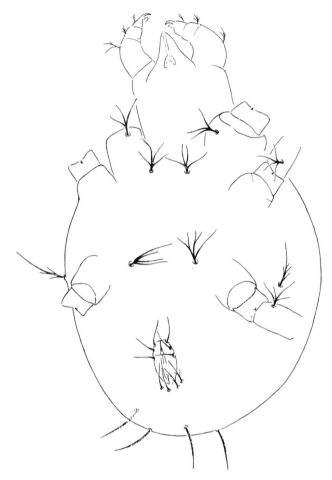

Figure 91, Venter of larva.
Neotrichobia arizonensis, new species

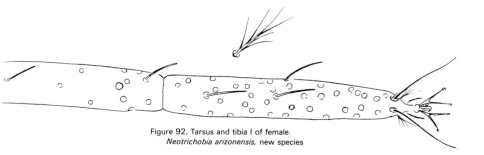

Figure 92, Tarsus and tibia I of female.
Neotrichobia arizonensis, new species

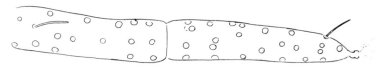

Figure 93, Tarsus and tibia II of female.
Neotrichobia arizonensis, new species

TETRANYCHINAE BERLESE

Tetranychini Berlese, 1913: 17.

Tetranychinae, Reck, 1950: 123; Pritchard and Baker, 1955: 96;
Wainstein, 1960: 145.

The Tetranychinae has three tribes, the Eurytetranychini, Tenuipalpoidini, and Tetranychini. Although these tribes are not as precisely defined as in the Bryobiinae, they are distinct enough to retain.

KEY TO THE GENERA AND TRIBES OF THE TETRANYCHINAE

1. Tarsus I with a single set, or without closely associated duplex setae, or duplex setae absent (Eurytetranychini) 2
   Tarsus I with two pairs of duplex setae, the proximal member of each pair shorter than the distal member; empodium clawlike or split distally 5

2. Empodial claw very large . . . . *Synonychus* Oudemans
   Empodial claw small or absent 3

3. Empodial claw small . . . . *Eurytetranychus* Oudemans
   Empodial claw absent 4

4. With two pairs of para-anal setae . . . . . *Eutetranychus* Banks
   With one pair of para-anal setae . . . . . *Aponychus* Rimando

5. Hysterosoma with fourth pair of dorsocentral setae in normal dorsal position; tarsus II with distal member of duplex setae long and tapering (Tetranychini) 6
   Hysterosoma with fourth pair of dorsocentral setae marginal; tarsus II with distal member of duplex setae a short solenidion (Tenuipalpoidini) . . . . . *Tenuipalpoides* Reck and Bagdasarian

6. With two pairs of para-anal setae 7
   With one pair of para-anal setae 16

7. Empodium clawlike 8
   Empodium ending in a tuft of hairs 13

8. Empodium a single clawlike structure 9
   Empodium split into two clawlike structures, usually with appendant hairs . . . . *Schizotetranychus* Trägårdh

9. Empodium without proximoventral hairs 10
   Empodium with proximoventral hairs 11

10. Dorsum of body with striae 12
    Dorsum of body reticulate . . . . *Mixonychus* Ryke and Meyer

11. Dorsal striae without spinules . . . . *Anatetranychus* Womersley
    Dorsal striae with spinules . . . . *Tylonychus* Miller

12. Empodial claw as long as or longer than proximoventral hairs which are at right angles to claw . . . . *Panonychus* Yokoyama
    Empodial claw shorter than proximoventral hairs which are less than at right angles to claw . . . . . *Allonychus* Pritchard and Baker
13. Hysterosomal striae transverse 14
    Hysterosomal striae longitudinal between third pair of dorsocentral setae
    . . . . *Mononychus* Wainstein
14. Striae normal; dorsal setae not set on tubercles; empodium split near middle 15
    Striae irregular in female, forming a basket weave pattern; setae set on strong tubercles; empodium split distally . *Neotetranychus* Trägårdh
15. Dorsal setae very short, not as long as intervals between their bases
    . . . . *Platytetranychus* Oudemans
    Dorsal setae as long as, or longer than, intervals between their bases
    . . . . *Eotetranychus* Oudemans
16. Empodium clawlike, with proximoventral hairs; duplex setae of tarsus I distal and approximate . . . . *Oligonychus* Berlese
    Empodium split distally, usually into three pairs of hairs; duplex setae of tarsus I well separated . . . . *Tetranychus* Dufour

EURYTETRANYCHINI RECK

Eurytetranychinae Reck, 1950: 123; Wainstein, 1960: 223.
Eurytetranychini, Pritchard and Baker, 1955: 100.

The tribe is characterized in that the duplex setae, if present, are not associated. There are four genera:

Eurytetranychus Oudemans
Eutetranychus Banks
Aponychus Rimando
Synonychus Miller

81

Eurytetranychus Oudemans

Eurytetranychus Oudemans, 1931: 224; Pritchard and Baker, 1955: 101; Wainstein, 1960: 223.

Type. (*Tetranychus latus*, Oudemans, not Canestrini and Fanago) = *Eurytetranychus buxi* (Garman).

The empodium is small and clawlike. The duplex setae are not closely associated with each other. The adults are large, globular and soft-bodied.

Members of the genus are not known from Arizona.

Eutetranychus Banks

Neotetranychus (Eutetranychus) Banks, 1917: 197.

Eutetranychus, McGregor, 1950: 267; Pritchard and Baker, 1955: 111; Wainstein, 1960: 226.

Type. Tetranychus banksi McGregor.

This genus is distinctive in that the empodium is lacking. Tarsus I bears a pair of loosely associated setae that may be homologous with the duplex setae.

Members of this genus have not been collected in Arizona.

Aponychus Rimando

Aponychus Rimando, 1966: 105.

Type. Aponychus corpuzae Rimando.

This genus is very similar to *Eutetranychus* except it has one pair instead of two pairs of anal setae in the female, and the dorsocentral setae IV are marginal. It is apparent *Eutetranychus shultzi, E. spinosus,* and *E. corderoi* belong in *Aponychus.*

In addition to the type species *Aponychus corpuzae,* a second species is described by Rimando, *A. rarus.*

We do not agree with Rimando's proposal of a new subfamily Aponychinae.

Rimando's publication was received by us after this manuscript was in press.

Synonychus Miller

Synonychus Miller, 1966: 57.

Type. Synonychus eucalypti Miller.

Like other genera of Eurytetranychini this genus lacks duplex setae on tarsi I and II. It differs from *Eurytetranychus* and *Eutetranychus,* in that the empodium is stout, clawlike, and nearly half as long as the tenent hairs of the true claw.

The only species for the genus occurs in Tasmania (Australia).

TENUIPALPOIDINI PRITCHARD AND BAKER

Tenuipalpoidini Pritchard and Baker, 1955: 97; Wainstein, 1960: 145.

The empodium is hooked and without ventral hairs, and the fourth pair of dorsocentral setae are marginal. There is only one genus: *Tenuipalpoides* Reck and Bagdasarian.

Tenuipalpoides Reck and Bagdasarian

Tenuipalpoides Reck and Bagdasarian, 1948: 183; Pritchard and Baker, 1955: 97; Wainstein, 1960: 147.

Type. Tenuipalpoides zizyphus Reck and Bagdasarian.

The genus has the characters of the tribe.

Tenuipalpoides dorychaeta Pritchard and Baker

Tenuipalpoides dorychaeta Pritchard and Baker, 1955: 99; Wainstein, 1960: 147.

This species differs from *Tenuipalpoides zizyphus* Reck and Bagdasarian; in having irregular rather than hooked peritremes, and the dorsal body setae are more elongate and tapering.

Tenuipalpoides dorychaeta has been collected in the crevices of small branches of black locust throughout the Central Eastern United States. A few collections have been made in the midwest. It has been collected in Arizona on New Mexican locust (*Robinia neomexicana* Gray) and snowberry (*Symphoricarpos palmeri* G. N. Jones).

TETRANYCHINI RECK

Tetranychinae Reck, 1950: 123.

Tetranychini, Pritchard and Baker, 1955: 124; Wainstein, 1960: 147.

This tribe is characterized by having two pairs of closely associated duplex setae on tarsus I, and a single pair on tarsus II. The fourth pair of dorsocentral setae are in the normal position. The empodium is either claw-like, or consists of several pairs of ventrally directed hairs. The following genera are here included:

Panonychus Yokoyama	*Platytetranychus* Oudemans
Allonychus Pritchard and Baker	*Anatetranychus* Womersley
Eotetranychus Oudemans	*Tylonychus* Miller
Schizotetranychus Trägårdh	*Mixonychus* Ryke and Meyer
Neotetranychus Trägårdh	*Oligonychus* Berlese
Mononychus Wainstein	*Tetranychus* Dufour

Panonychus Yokoyama

Panonychus Yokoyama, 1929: 531.
Metatetranychus Oudemans, 1931: 199; Pritchard and Baker, 1955: 127.
   Type. (*Panonychus mori* Yokoyama) = *Panonychus citri* (McGregor).
   There are two pairs of para-anal setae. The empodium is clawlike and bears three pairs of proximoventral setae. The dorsal body setae are borne on strong tubercles.
   There are two well known species, *Panonychus citri* (McGregor) and *P. ulmi* (Koch). Only the citrus red mite, *P. citri,* is known to occur in Arizona.

Panonychus citri (McGregor)

Tetranychus citri McGregor, 1916: 284.
Paratetranychus citri, McGregor, 1919: 672.
Metatetranychus citri, Pritchard and Baker, 1955: 133.
Panonychus citri, Ehara, 1956: 500.
   The adults are reddish or purple with reddish setae borne on strong tubercles. They are similar to *Panonychus ulmi* (Koch), European red mite, but differ by having the outer sacrals and the clunals all similar in length, about one-third the length of the inner sacrals. In *P. ulmi* the outer sacrals are about two-thirds as long as the inner sacrals and the clunals about one-third as long. The dorsal setae of *P. ulmi* are white.
   Panonychus citri was discovered in Arizona at Yuma on May 4, 1967, by Ted Townsend and determined by D. M. Tuttle. It was found on various citrus including: lemon, orange, tangerine, citron, and grapefruit.

Allonychus Pritchard and Baker

Allonychus Pritchard and Baker, 1955: 137; Baker and Pritchard, (1962) 1963: 312; Wainstein, 1960: 200.
   Type. *Septanychus braziliensis* McGregor.
   The mites of this genus possess two pairs of para-anal setae. The empodium consists of a large mediodorsal spur with three pairs of proximoventral hairs dissimilar in length and set at an angle of less than 45° with the spur. The palpal claw is bifurcate. The lobes of the dorsal striae are much taller than broad, and pointed distally.
   These mites are tropical in distribution, having been found in Mexico, Central America, and South America. They have not been collected in Arizona.

Eotetranychus Oudemans

Eotetranychus Oudemans, 1931: 224; Pritchard and Baker, 1955: 138.

Schizotetranychus (*Eotetranychus*), Wainstein, 1960: 178.

Type. *Trombidium tiliarum* Hermann.

The genus *Eotetranychus* is closely allied to *Schizotetranychus* and has been synonymized as a subgenus by Wainstein (1960). Because of host plant relations and some morphological differences, we are here retaining the two genera separate.

There are two pairs of para-anal setae; the duplex setae are distal and adjacent on tarsus I; the empodium is split into three pairs of ventrally directed hairs, usually of equal strength. The striae bear small lobes, are longitudinal on the propodosoma and transverse on the hysterosoma. In general, these mites feed on the undersurface of broadleaf plants.

KEY TO THE SPECIES OF EOTETRANYCHUS IN ARIZONA (FEMALES)

1. Striae anterior to female genital area not transverse 2
   Striae anterior to female genital area transverse 6
2. Striae anterior to genital area longitudinal 3
   Striae anterior to genital area irregularly longitudinal 4
3. Genital flap with only transverse striae . . . *cercocarpi,* new species
   Genital flap with longitudinal striae . . . . . *sexmaculatus* (Riley)
4. Peritreme hooked distally 5
   Peritreme ending in a simple bulb distally . . . *gambelii,* new species
5. Solenidia of tarsi II-IV shorter than width of tarsi . . . .
   . . . . *salix* Tuttle and Baker
   Solenidia of tarsi II-IV much longer than width of tarsi . . . . .
   . . . . *fallugiae,* new species
6. Tibia II with seven or eight tactile setae 7
   Tibia II with five tactile setae . . . . *potentillae* Tuttle and Baker
7. Tibia II with eight tactile setae 8
   Tibia II with seven tactile setae . . . . *prosopis* Tuttle and Baker
8. Peritreme hooked or anastomosing distally 8
   Peritreme straight distally . . . . *deflexus* (McGregor)
9. Peritreme hooked distally 9
   Peritreme anastomosing distally . . . . *fremonti* Tuttle and Baker
10. Tarsus I with five tactile setae proximal to duplex setae 10
    Tarsus I with four tactile setae proximal to duplex setae . . . . .
    . . . . *mastichi* DeLeon
11. Longitudinal striae of propodosoma simple 11
    Longitudinal striae of propodosoma irregular . . *frosti* (McGregor)

12. The females of the following species are similar and males must be studied for specific determinations:

malvastris (McGregor) . . . aedeagus almost triangular.

weldoni (Ewing) . . . . . aedeagus long, strong and only slightly sinuous.

lewisi (McGregor) . . . . aedeagus downturned and slightly sigmoid, with dorsal angle.

yumensis (McGregor) . . . aedeagus short, upturned and with slight sigmoid shape.

juniperus Tuttle and Baker, (female not known) . . . . aedeagus short, obtusely downturned and sigmoid.

Eotetranychus cercocarpi, new species

(Fig. 94)

Although superficially resembling *Eotetranychus lewisi* (McGregor), this species is distinctive in having longitudinal striae anterior to the genital flap of the female which possesses transverse striae, and the peritreme ends in a simple bulb.

Male. Terminal sensillum of palpus minute, about as long as broad. Peritremes ending in a simple bulb distally. Tarsus I with three tactile setae and three solenidia posterior to the duplex setae; tibia I with eight tactile setae and three solenidia. A short proximal solenidion on tarsi III and IV. Empodium of tarsus I clawlike, double; other empodia normal. Aedeagus downcurved, tapering to tip which is slightly sigmoid. Length of body including rostrum 320 μ.

Female. Terminal sensillum of palpus about one and one-half times as long as broad. Peritremes ending in a simple bulb distally. Tarsus I with five tactile setae and one solenidion proximal to duplex setae. Tibia I with nine tactile setae and one solenidion. Tibia II with eight tactile setae. Striae of genital flap transverse; striae anterior to genital flap longitudinal. Length of body including rostrum 345 μ.

Holotype. Male, *ex Cercocarpus montanus* Raf., McNary, Arizona, February 20, 1965, by D. M. T.

Paratypes. One male and seven females with the above data.

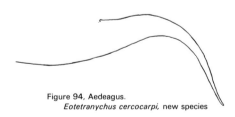

Figure 94, Aedeagus.
Eotetranychus cercocarpi, new species

Eotetranychus sexmaculatus (Riley)

Tetranychus 6-maculatus Riley, 1890: 225.
Eotetranychus sexmaculatus, Tuttle and Baker, 1964: 21.

In *Eotetranychus sexmaculatus,* the striae anterior to the genital region of the female are longitudinal, as are the striae on the anterior region of the genital flap. The aedeagus is long, slender, curving slightly dorsad near the middle of the shaft and caudoventrally directed distally.

This species is primarily a pest of citrus in California and Florida and has been found on the same host in Arizona.

Eotetranychus gambelii, new species

(Fig. 95)

Eotetranychus gambelii keys out to *E. perplexus* (McGregor) in Pritchard and Baker (1955), but the female has irregularly longitudinal striae rather than transverse striae anterior to the genital flap, and the aedeagus is attenuated distally rather than expanded.

Male. Terminal sensillum of palpus minute, not much longer than broad. Peritremes ending distally in a simple bulb. Tarsus I with four tactile setae and three solenidia proximal to the duplex setae; tibia I with nine tactile setae and one solenidion. A short solenidion each on tarsi III and IV. Empodium of tarsus I clawlike, double; other empodia normal. Aedeagus downcurved, tapering but strong, and sigmoid. Length of body including rostrum 332 μ.

Female. Terminal sensillum of palpus about two and one-half times as long as broad. Peritremes ending in a simple bulb distally. Tarsus I with four tactile setae and one solenidion proximal to duplex setae; tibia I with nine tactile setae and one solenidion. Tibia II with eight setae. Genital flap with transverse striae; area anterior to flap with irregular longitudinal striae as in *Eotetranychus pallidus* (Garman) (Pritchard and Baker, 1955: 139). Length of body including rostrum 351 μ.

Holotype. Male, *ex Quercus gambelii* Nutt., McNary, Arizona, February 20, 1965, by D. M. T.

Paratypes. Two males and three females with the above data.

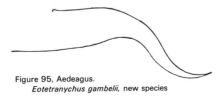

Figure 95, Aedeagus.
Eotetranychus gambelii, new species

87

Eotetranychus salix Tuttle and Baker

Eotetranychus salix Tuttle and Baker, 1964: 24.

This species, known only from the female, will key out to *Eotetranychus clitus* Pritchard and Baker (1955) but differs in having four instead of five proximal setae on tarsus I, and in having a much shorter solenidion on tibia I, and tarsi III and IV.

Eotetranychus salix is known only from Arizona on *Salix gooddingii* Ball.

Eotetranychus fallugiae, new species

(Figs. 96-97)

The male and female both key out to *Eotetranychus clitus* Pritchard and Baker (1955), but the male is distinctive in that the terminal sensillum of the palpus is strongly reduced, and that of the female is slightly longer than broad; the aedeagus is also longer and more slender.

Male. Terminal sensillum of palpus tiny; stylophore long and slightly indentate anteriorly; peritreme hooked distally. Dorsal body setae longer than intervals between their bases. Tarsus I with two slender solenidia, tibia I with nine tactile setae and two solenidia; genu I with five tactile setae; femur I with ten tactile setae. Tarsus II with one slender solenidion; tibia II with six tactile setae; genu II with four tactile setae; femur II with seven tactile setae. Tarsus III with long slender proximal solenidion; tibia III with five tactile setae; genu III with four tactile setae; femur III with four tactile setae. Leg IV similar to leg III except femur with only three setae. Empodium I clawlike; other empodia with one pair of strong ventrally directed hairs and two weaker pairs. Aedeagus slender, downcurved and broadly sigmoid. Length of body 255 μ; including rostrum 319 μ.

Female. Terminal sensillum of palpus short and broadly rounded, not much longer than broad; stylophore broad, distinctly indentate anteriorly; peritreme hooked distally. Body setae longer than intervals between their bases. Genital flap with transverse striae; area anterior to flap with longitudinal striae. Tarsus I with slender solenidion of medium length; tibia I with eight tactile setae and one slender solenidion; genu I with five tactile setae; femur I with ten tactile setae. Tarsus III with one slender solenidion; tibia II with seven tactile setae; genu II with five setae; femur II with seven setae. Tarsus III with long slender proximal solenidion; tibia III with five tactile setae; genu III with four tactile setae; femur II with four tactile setae. Leg IV similar to III except femur with only three tactile setae. Length of body 293 μ; including rostrum 344 μ.

Holotype. Male, *ex Fallugia paradoxa* (D. Don) Endl., Portal, Arizona, August 28, 1964, by D. M. T.

Paratypes. Three males and six females with the above data.

88

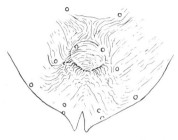

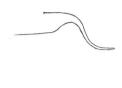

Figure 96, Genital-ventral region of female.
Eotetranychus fallugiae, new species

Figure 97, Aedeagus.
Eotetranychus fallugiae, new species

Eotetranychus potentillae Tuttle and Baker

Eotetranychus potentillae Tuttle and Baker, 1964: 32.

This species is distinct in having the tarsal sensory setae short and dorsal body setae long and pubescent in both sexes. The aedeagus is upturned and sigmoid.

The description of the female was unintentionally omitted in the original description and is given here.

Female. Terminal sensillum of palpus not strong, about two times as long as wide; other sensilla as in a male. Peritreme hooked distally. Tibia I with eight tactile and one short sensory setae; tarsus I with two tactile and one sensory setae proximal to duplex setae. Tibia II-IV each with five tactile setae. Tarsus II with two tactile and one very short sensory setae proximal to duplex setae. Tarsus II and IV each with a short sensory seta and eight tactile setae. Genital striae transverse. Dorsal body setae very long and pubescent. Length of body 465 μ.

Eotetranychus potentillae is known only from Arizona on *Potentilla hippiana* Lehm.

Eotetranychus prosopis Tuttle and Baker

Eotetranychus prosopis Tuttle and Baker, 1964: 24.

This species is similar to *Eotetranychus clitus* Pritchard and Baker (1955), but differs in having the angle of the aedeagus more obtuse posteriorly, and in that the terminal sensillum of the palpus is small, conical, and about as long as wide.

This species has been found in Arizona on *Prosopis juliflora* (Swartz) DC.

89

Eotetranychus deflexus (McGregor)

Tetranychus deflexus McGregor, 1950: 284.
Eotetranychus deflexus Pritchard and Baker, 1955: 206.

The aedeagus of *Eotetranychus deflexus* (McGregor) is strongly bent ventrad near the middle with the bent portion tapering and sigmoid. The terminal sensillum of the palpus is rudimentary as in *Eotetranychus prosopis*.

The species was originally described from snowberry (*Symphoricarpos* sp.) in Oregon. It occurs on *Symphoricarpos oreophilus* Gray and *S. palmeri* G. N. Jones in Arizona.

Eotetranychus fremonti Tuttle and Baker

Eotetranychus fremonti Tuttle and Baker, 1964: 260.

The terminal sensillum of the male is small, conical and about as long as wide. The peritreme of the male is hooked and that of the female is slightly anastomosing. The aedeagus is bent ventrad, with a short anterior and a much longer angulation.

This species has been collected in Arizona on *Populus fremontii* S. Wats.

Eotetranychus mastichi DeLeon

Eotetranychus mastichi DeLeon, 1957: 111.
Eotetranychus oistus Beer and Lang, 1958: 1241.

This species is distinctive in having the aedeagus expanded distally, similar to that of *Eotetranychus perplexus* (McGregor), but the posterior tip turns ventrad rather than dorsad. The terminal sensillum of the male palpus is small and slightly longer than broad. Empodium I of the male is split into two equal claws; the other empodia are normal. The peritreme is enlarged distally. Tarsus I has four tactile setae and three solenidia proximal to the duplex setae; Tibia I possesses nine tactile setae and four solenidia. Tarsus II of the male possesses a strong solenidion. Tibia II possesses eight tactile setae. Tarsi III and IV each with a single solenidion. In the female, the terminal sensillum of the palpus is about two times as long as broad. The peritreme ends in a hook. The stylophore is either emarginate or rounded anteriorly. The striae of the genital flap are transverse, as are those anterior ot this region. Tarsus I has four tactile setae and one solenidion proximal to the duplex setae; tibia I possesses nine tactile setae and one solenidion. Tarsus II has three tactile setae and one short solenidion proximal to the duplex setae; there are eight tactile setae on tibia II. Tarsi III and IV each possesses a short dorsal solenidion.

This species was originally described from *Sideroxylon foetidissima* Jacq., Coral Gables, Florida. It has been collected in Arizona on *Cowania mexicana* D. Don. var. *stansburiana* (Torr.) Jepson, Prescott, August 18, 1965, by D. M. T. It has also been collected in Mexico and Nicaragua.

90

Eotetranychus frosti (McGregor)

Tetranychus frosti McGregor, 1952: 142.
Eotetranychus frosti, Tuttle and Baker, 1964: 21.

This species is distinctive in having irregular striae on the propodosoma of the female, a hooked peritreme, and in that the aedeagus is upturned and sigmoid.

Eotetranychus frosti has been collected on rose (*Rosa* spp.) in Arizona. It has also been found in California, Louisiana, Missouri, North Dakota, and Ohio.

Eotetranychus malvastris (McGregor)

Tetranychus malvastris McGregor, 1950: 290.
Eotetranychus malvastris, Tuttle and Baker, 1964: 20.

The female of *Eotetranychus malvastris* (McGregor) is similar to several other species from southwestern United States, and the species can be recognized only by the shape of the aedeagus, which is short and almost triangular.

In Arizona this species occurs on *Cassia* sp., *Cercidium microphyllum* (Torr.) Rose and Johnston, *Croton californicus* Muell. Arg., *Ditaxis lanceolata* (Benth.) Pax and Hoffmann, *Phacelia crenulata* Torr., *Sida hederacea* (Dougl.) Torr., *Sphaeralcea ambigua* Gray, and *Sphaeralcea orcuttii* Rose. It has also been collected in California.

Eotetranychus weldoni (Ewing)

Tetranychus weldoni Ewing, 1913: 457.
Eotetranychus weldoni, Tuttle and Baker, 1964: 20.

The female of *Eotetranychus weldoni* (Ewing) is similar to females of several other species of *Eotetranychus* found in the southwestern United States. The male is necessary for specific recognition. The peritreme is hooked distally, and the aedeagus is strong and only slightly sigmoid.

This species has been found in Arizona on willows *(Salix)* and poplars *(Populus).* It has also been collected in California, Kansas, and Colorado.

Eotetranychus lewisi (McGregor)

Tetranychus lewisi McGregor, 1943: 127.
Eotetranychus lewisi, Tuttle and Baker, 1964: 21.

The female of *Eotetranychus lewisi* (McGregor) is similar to females of several other species of *Eotetranychus* in the southwestern United States, and the species must be identified by the male aedeagus. The aedeagus is relatively small and short, bending ventrad but with a slightly upturned posterior tip.

This species has been collected in Arizona on several species of plants. It has also been found throughout Mexico and Central America, as well as in California and in greenhouses in Washington state.

Tetranychus yumensis McGregor, 1934: 256.

Eotetranychus yumensis, Tuttle and Baker, 1964: 20.

Eotetranychus yumensis (McGregor) is also another species in which the female cannot be separated from several other females of different species. The male is distinctive in that the aedeagus forms a shallow, nearly sigmoid dorsal curve, the distal end tapering gradually to a point.

This species has been taken on *Citrus* spp. in Arizona as well as in California. It has also been taken on other hosts in Arizona: *Oenothera clavaeformis* Torr. and Frém., *Ricinus communis* L., *Rosa multiflora* Thunb., *Sorghum vulgare* Pers., and *Vitis vinifera* L.

Eotetranychus juniperus Tuttle and Baker

Eotetranychus juniperus Tuttle and Baker, 1964: 30.

This species is known only from the male in which the aedeagus is downturned and obtusely sigmoid, and the terminal sensillum of the palpus is rudimentary.

Eotetranychus juniperus has been collected on *Juniperus deppeana* Steud. in Arizona.

Schizotetranychus Tragardh

Schizotetranychus Trägårdh, 1915: 277; Pritchard and Baker, 1955: 225; Wainstein, 1960: 166.

Type. Tetranychus schizopus Zacher.

There are two pairs of para-anal setae; the duplex setae are distal and approximate to each other on tarsus I; the empodia are strong, split and clawlike; all three pairs of empodial hairs may or may not be present. The striae on the propodosoma may be longitudinal, or may be transverse in part or in whole; that of the hysterosoma is usually transverse, but may be longitudinal on the anterior portion; striae anterior to the female genitalia may either be transverse or longitudinal. In general, these mites are grass feeders.

KEY TO THE SPECIES OF SCHIZOTETRANYCHUS IN ARIZONA (FEMALES)

1. First pair of dorsocentral hysterosomal setae one-half as long as first pair of dorsolateral setae 2
   First pair of dorsocentral setae as long as or nearly as long as first pair of dorsocentral setae 3
2. First four pairs of dorsocentral setae about one-half as long as dorsolateral setae . . . . *celtidis,* new species
   First two pairs of dorsocentral setae about one-half as long as dorsolateral setae . . . . *eremophilus* McGregor

Schizotetranychus celtidis, new species

(Figs. 98-99)

This species is distinctive in having all first four pairs of the dorsocentral hysterosomal setae subequal in length and about one-half as long as the dorsolateral setae. It is known only from the female.

Female. Rostrum elongate; terminal sensillum of palpus about two and one-half times longer than broad, stylophore rounded anteriorly; peritreme expanded into simple bulb distally. Striae transverse on dorsocentral portion of propodosoma; striae transverse on hysterosoma. Third pair of propodosomal setae longer than first and second pairs; first three pairs of dorsocentral hysterosomal setae short, not as long as distance between their bases; fourth pair of dorsolateral setae slightly longer; all dorsocentral setae shorter than dorsolateral setae; humeral setae longer than others. Empodium with a pair of dorsal hairs on each clawlike section. Tarsus I with long, slender solenidion — as long as segment — and with two tactile setae proximal to duplex setae; tibia I with seven tactile setae and one straight stout solenidion about as long as width of segment; genu I with five setae; femur I with seven tactile setae. Tarsus II with one dorsal proximal seta; tibia II with five tactile setae; genu II with five tactile setae; femur II with five tactile setae. Tarsus III

93

with dorsomedian solenidion longer than segment; tibia III with five tactile setae; genu III with three tactile setae; femur III with three tactile setae. Leg IV similar to leg III. Striae of genital flap and in area anterior to flap transverse. Length of body 225μ; including rostrum 306 μ.

Male. Not known.

Holotype. Female, *ex Celtis reticulata* Torr., Portal, Arizona, August 28, 1964, by D. M. T.

Paratypes. Two females with the above data.

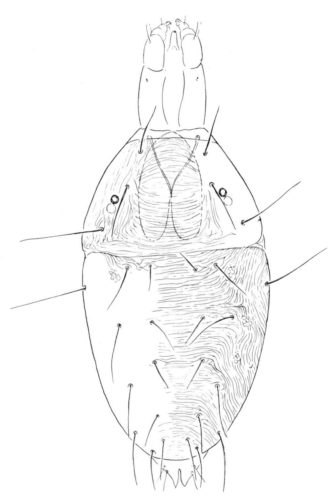

Figure 98, Dorsum of female.
Schizotetranychus celtidis, new species

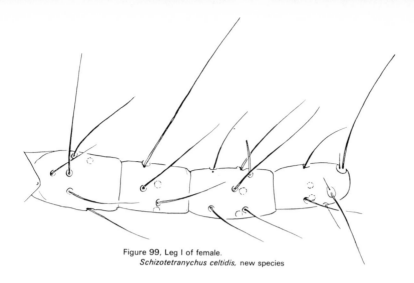

Figure 99, Leg I of female.
Schizotetranychus celtidis, new species

Schizotetranychus eremophilus McGregor

Schizotetranychus eremophilus McGregor, 1950: 311; Tuttle and Baker, 1964: 32.

Schizotetranychus eremophilus is distinctive in that the first pair of dorsocentral hysterosomal setae are shorter than the others, as well as being much shorter than the first pair of dorsocentral setae. The striae are transverse on the genital flap and on the area anterior to the flap; the striae are longitudinal on the propodosoma and transverse on the hysterosoma.

This species has been taken in Arizona on *Aristida glabrata* (Vasey) Hitchc., *Bouteloua barbata* Lag., *Cynodon dactylon* (L.) Pers., *Distichlis stricta* (Torr.) Rydb., and *Tridens pulchellus* (H.B.K.) Hitchc.

Schizotetranychus nugax Pritchard and Baker

Schizotetranychus nugax Pritchard and Baker, 1955: 264.

The first three pairs of dorsocentral hysterosomal setae are much shorter than the distance between their bases; the fourth and fifth pairs are much longer than the others; the first pair of dorsolateral setae are subequal in length to the first pair of dorsocentral setae; other dorsolateral and humeral setae are equal in length to the fourth and fifth pairs of dorsocentral setae. The striae of the propodosoma are longitudinal only to the second pair of propodosomal setae; striae on the anterior portion of the genital flap are longitudinal, as are the striae in the area anterior to the genitalia.

This species was described from females collected on native grasses, Portales, New Mexico, but is here included because of the probability of also being found in Arizona within a short time.

95

Schizotetranychus boutelouae, new species
(Figs. 100-101)

This species is closely allied to *Schizotetranychus nugax* Pritchard and Baker (1955), differing principally in having U-like striae on the propodosoma rather than longitudinal striae.

Female. Body elongate. Rostrum elongate, rounded anterior; terminal sensillum of palpus about one and one-half times as long as broad, and bluntly pointed distally; petritreme simple, expanded distally. Striae of propodosoma U-shaped, longitudinal only to the second pair of propodosomal setae; striae transverse on hysterosoma. First row of hysterosoma setae short and of equal length; in second row dorsocentrals slightly shorter; third row similar; fourth and fifth row of dorsocentrals longer and subequal in length. Genital flap with transverse striae somewhat longitudinal in anterior central area; area anterior to flap with longitudinal striae. All empodial claws split and strong. Tarsus I with one long slender solenidion and one tactile seta proximal to duplex setae; tibia I with seven tactile and one solenidion. Tarsus II with a single solenidion; tibia II with five tactile setae; tarsi III and IV each with a long slender solenidion. Length of body 383 μ; including rostrum 480 μ.

Male. Dorsal setal and striation pattern similar to that of the female with striae longitudinal between second pair of propodosomal setae. Aedeagus upturned. Stylophore slightly indentate anteriorly; peritreme bent and with small bulb distally. Tarsus I with three tactile setae and three slender solenidia proximal to duplex setae; tibia I with seven tactile setae and two solenidia; genu I with five tactile setae; femur I with seven tactile setae. Tarsus II with one solenidion; tibia II with five tactile setae; genu II with three tactile setae; femur II with five tactile setae. Tarsus III with dorsal solenidion longer than segment; tibia III with five tactile setae; genu III with three tactile setae; femur III with three tactile setae. Leg IV similar. Aedeagus seen only dorsally but probably ventrally directed and sigmoid as in other species. Length of body 268 μ; including rostrum 370 μ.

Holotype. Female, *ex Bouteloua rothrockii* Vasey, Oracle, Arizona, July 26, 1964, by D. M. T.

Paratypes. One male and eleven females with the above data.

96

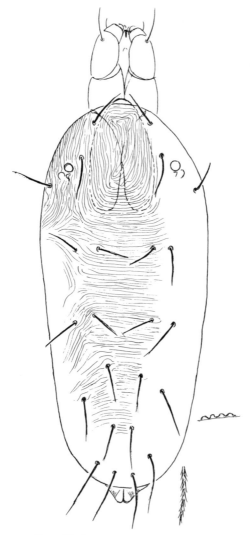

Figure 100, Dorsum of female.
Schizotetranychus boutelouae, new species

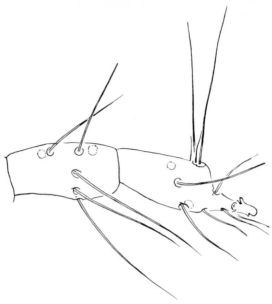

Figure 101, Tarsus and tibia I of female.
Schizotetranychus boutelouae, new species

Schizotetranychus fluvialis McGregor

Schizotetranychus fluvialis McGregor, 1928: 13; Pritchard and Baker, 1955: 254.

This species is distinctive in having the dorsal setae all very short, much shorter than the intervals between their bases. The striae of the propodosoma are longitudinal dorsally to a level with the second pair of propodosomal setae; the striae of the hysterosoma are transverse; the striae of the genital flap are transverse, those of the area anterior to the flap irregularly longitudinal. The aedeagus is upturned and sigmoid. The tarsi are blunt distally.

Schizotetranychus fluvialis was originally collected from grass, *Muhlenbergia rigens* (Benth.) Hitchc., in California. In Arizona, specimens were taken on *Aristida adscensionis* L., Marana, April 2, 1963, and on *Muhlenbergia rigens* (Benth.) Hitchc., Prescott, August 19, 1965, by D. M. T.

Schizotetranychus montanae, new species

(Figs. 102-104)

The dorsal body setae are about as long as the intervals between their bases and subequal in length. *Schizotetranychus montanae* is closely related to *S. lycurus* Tuttle and Baker, but it can be easily separated by the very long dorsal seta on tibia IV.

Female. Rostrum elongate; terminal sensillum of palpus about two times as long as broad, somewhat sharpened distally; stylophore rounded anteriorly; peritreme expanded distally. Striae longitudinal on propodosoma and transverse on hysterosoma; striae transverse on genital flap and irregularly longitudinal in area anterior to flap. Propodosomal setae subequal in length, the second pair not reaching the base of the third pair; dorsal hysterosomal setae subequal in length, not reaching to base of next pair except for fourth and fifth transverse rows; humeral setae about as long as posterior setae. Empodial claws strong, solid. Tarsus I with long slender solenidion and three tactile setae proximal to duplex setae; tibia I with seven tactile setae and one slender solenidion; genu I with five tactile setae; femur I with nine tactile setae, the dorsomedian seta reaching to tarsus. Tarsus III with dorsoproximal solenidion longer than segment; tibia III with five tactile setae; genu III with three tactile setae; femur III with three tactile setae. Leg IV similar to leg III. Length of body 325 μ; including rostrum 408 μ.

Male. Not known.

Holotype. Female, *ex Muhlenbergia montanae* (Nutt.) Hitchc., McNary, Arizona, August 15, 1963, by D. M. T.

Paratypes. Seven females with the above data.

Schizotetranychus lycurus Tuttle and Baker

Schizotetranychus lycurus Tuttle and Baker, 1964: 34.

The dorsal setae are shorter than the intervals between their bases and are subequal in length. The tarsi are not blunt distally, and there are three tactile setae proximal to the duplex setae on tarsus I. It differs from *Schizotetranychus montanae*, new species, in that the solenidia on tarsi III and IV are not longer than the segments. The body striae are longitudinal on the propodosoma and transverse on the hysterosoma; the striae are longitudinal in the area anterior to the female genitalia, and also longitudinal on the anterior portion of the genital flap.

Schizotetranychus lycurus has been collected in Arizona on *Lycurus phleoides* H.B.K.

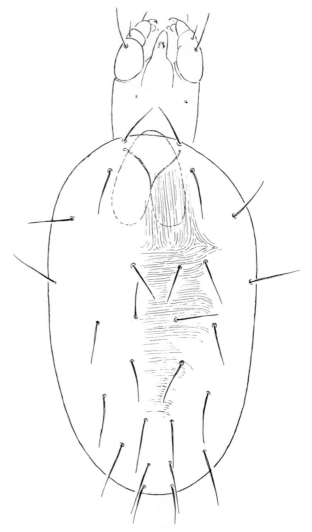

Figure 102, Dorsum of female.
Schizotetranychus montanae, new species

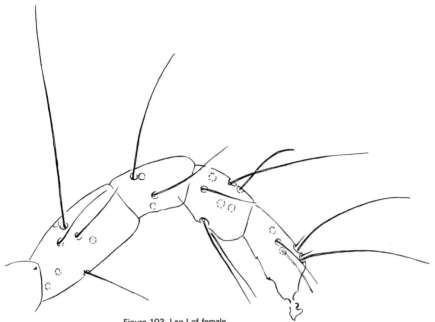

Figure 103, Leg I of female.
Schizotetranychus montanae, new species

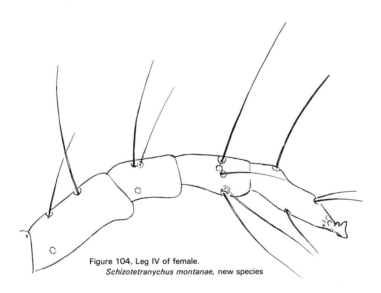

Figure 104, Leg IV of female.
Schizotetranychus montanae, new species

Schizotetranychus cynodonis McGregor

Schizotetranychus cynodonis McGregor, 1950: 309; Tuttle and Baker, 1964: 32.

The male of this species has a pair of triangular genital stylets that accompany the aedeagus. The female has strongly pubescent setae on the venter of the tibiae and tarsi; the dorsal setae of the female are subequal in length and longer than the intervals between their bases. The striae on the propodosoma of the female are longitudinal; those on the hysterosoma transverse; the striae of the genital flap and the area anterior are transverse.

This species has been collected in Arizona on *Distichlis stricta* (Torr.) Rydb. It has also been taken in California on *Cynodon dactylon* (L.) Pers. and *Distichlis stricta* (Torr.) Rydb.

Schizotetranychus hilariae, new species

(Figs. 105-107)

This species is similar to *Schizotetranychus elymus* McGregor in the length of the dorsal body setae, but differs in having longitudinal striae on the propodosoma.

Female. Body elongate; rostrum long, reaching to the middle of tibia I; terminal sensillum of palpus about two times as long as broad; peritremes hooked or slightly hooked distally. Striae longitudinal on propodosoma; in general transverse on hysterosoma. All dorsal body setae slender, longer than intervals between bases, and subequal in length. Genital flap and area anterior to flap with transverse triae. All empodia split with two weak dorsal members and strong ventral spur. Tarsus I with three tactile setae; tibia I with eight tactile setae and one solenidion. Tarsus II without solenidion; tibia II with five tactile setae. Both tarsi III and IV with a long proximal solenidion. Length of body 255 μ; width 364 μ.

Male. In poor condition. Empodial claws split, with two weak dorsal members and strong ventral spur. Aedeagus distorted, upturned, apparently knobbed as figured. Tarsus I with one tactile and three solenidia proximal to duplex setae; tibia I with eight tactile setae and two solenidia. Tarsus II with two tactile setae and one solenidion proximal to duplex setae; tibia II with five tactile setae. A long slender proximal solenidion on tarsi III and IV. Badly smashed and not measurable.

Holotype. Female, *ex Hilaria rigida* (Thurb.) Benth., Dateland, Arizona, August 19, 1964, by D. M. T.

Paratypes. One male, *ex Hilaria rigida* (Thurb.) Benth., Gila Bend, Arizona, October 15, 1963, by D. M. T.; two females from the above species,

Yuma, Arizona, April 10, 1960, and June 20, 1962, by D. M. T., two females from the above host, Gila Bend, Arizona, October 15, 1963, by D. M. T., one female from the above host, Salome, Arizona, June 9, 1965, by D. M. T.

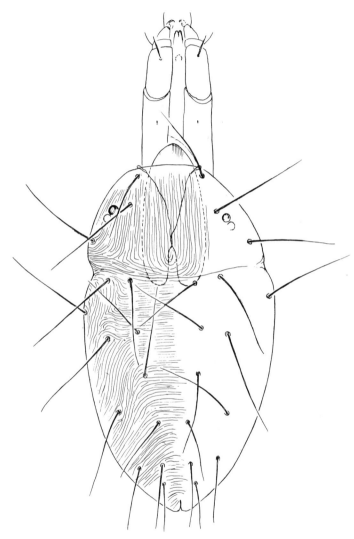

Figure 105, Dorsum of female.
Schizotetranychus hilariae, new species

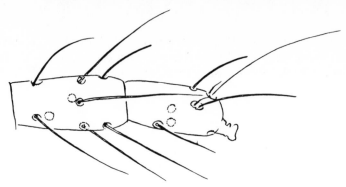

Figure 106, Tarsus and tibia I of female.
Schizotetranychus hilariae, new species

Figure 107, Aedeagus.
Schizotetranychus hilariae, new species

Schizotetranychus elymus McGregor

Schizotetranychus elymus McGregor, 1950: 310; Tuttle and Baker, 1964: 34.

The male of this species has a dorsally directed sigmoid aedeagus. The dorsal body setae of both sexes are longer than the distances between their bases. The striations of the female are longitudinal only on the anterior portion of the propodosoma between the first and second pairs of propodosomal setae; the striae of the rest of the dorsocentral portion of the propodosoma and hysterosoma are transverse; the striae of the genital flap and the area anterior to this are transverse.

Schizotetranychus elymus has been taken on grasses in California and Utah. It has been collected in Arizona on *Agropyron trachycaulum* (Link) Malte., *Aristida adscensionis* L., *Bouteloua hirsuta* Lag., *Distichlis stricta* (Torr.) Rydb., *Malva parvifloria* L., *Tridens pulchellus* (H.B.K.) Hitchc., *Typha latifolia* L., and *Vicia pulchella* H.B.K.

Neotetranychus Tragardh

Neotetranychus Trägårdh, 1915: 23; Pritchard and Baker, 1955: 215; Wainstein, 1960: 198.

Type. Neotetranychus rubi Trägårdh.

There are two pairs of para-anal setae. The empodia are split distally in both sexes. The dorsal body setae are long and set on prominent tubercles.

104

The striae of the female are transverse on the hysterosoma and have a basket weave or similar pattern.

At present, only the type species and another being described from Mexico belong in this genus. Those listed in Pritchard and Baker (1955), other than the type, belong elsewhere. This genus is not known to occur in Arizona.

Mononychus Wainstein

Mononychus Wainstein, 1960: 198.

Type. Tetranychus planki McGregor.

This genus contains those mites with two pairs of para-anal setae, distal adjacent duplex setae on tarsus I, a split empodium, and longitudinal striae between the third pair of dorsocentral setae. The lobes of the striae are prominent, there may be an anastomosing striation pattern, and the setae may be borne on weak tubercles.

Only one species is known from Arizona, *Mononychus siccus* (Pritchard and Baker).

Mononychus siccus (Pritchard and Baker), new combination

Neotetranychus siccus Pritchard and Baker, 1955: 219; Wainstein 1960: 198.
Eotetranychus siccus, Tuttle and Baker, 1964: 26.

This species has long, slender dorsal setae set on weak tubercles. The striae are longitudinal between the third pair of dorsocentral hysterosomal setae. The stylophore is about twice as long as broad; the peritreme is simple distally. Tarsus I has four tactile setae and one solenidion proximal to the duplex setae; tibia I has seven tactile setae and one solenidion.

Mononychus siccus was first collected on ironwood, *Olneya tesota* Gray, at Phoenix, Arizona. Since then several collections have been made in the desert areas of Arizona by D. M. T.

Platytetranychus Oudemans

Platytetranychus Oudemans, 1931: 224; Pritchard and Baker, 1955: 138, 163; Wainstein, 1960: 228.

Type. Tetranychus gibbosus Canestrini.

The dorsal body setae are much shorter than the distance between their bases; the striae are longitudinal on the propodosoma and transverse on the hysterosoma; the peritreme ends in a simple bulb; tibia II has five tactile setae; and the duplex setae of tarsus I are distal and adjacent.

The genus contains *Platytetranychus gibbosus* (Canestrini), *P. libocedri* (McGregor), and *P. thujae* (McGregor), all from conifers. The new species,

P. symphoricarpi, described here was collected on *Symphoricarpos oreophilus* Gray, but the mite-host relationship may be accidental.

KEY TO THE SPECIES OF PLATYTETRANYCHUS IN ARIZONA (FEMALES)

1. Dorsocentral setae tapered distally; tibia I with nine tactile setae and one solenidion 2
   Dorsocentral setae rounded distally; tibia I with seven tactile setae and one solenidion . . . . *thujae* (McGregor)
2. Terminal sensillum of palpus three times as long as wide; lobes on dorsal striae triangular . . . . *libocedri* (McGregor)
   Terminal sensillum of palpus less than two times as long as wide; lobes on dorsal striae broadly rounded . . . . *symphoricarpi,* new series

Platytetranychus thujae (McGregor), new combination

Tetranychus thujae McGregor, 1950: 303.
Eotetranychus thujae, Pritchard and Baker, 1955: 159.

This species is similar to *Platytetranychus libocedri* in having short dorsal setae. It differs in that the dorsocentral setae are shorter and rounded distally and tibia I possesses seven tactile setae and a solenidion. Tarsus I of the female may have three or four tactile setae proximal to the duplex setae; tarsus I of male has three tactile setae and a solenidion proximal to the duplex setae. The aedeagus is distinctive in that it narrows abruptly near the base of the shaft, terminates in a fingerlike end, and is broadly rounded at the tip.

Platytetranychus thujae has been collected on *Juniperus osteosperma* (Torr.) Little, McNary, Arizona, August 11, 1963, by D. M. T. It is common in eastern, southeastern, and midwestern United States on cupressaceous conifers.

Platytetranychus libocedri (McGregor), new combination

Tetranychus libocedri McGregor, 1936: 771.
Eotetranychus libocedri, Tuttle and Baker, 1964: 20.

In this species, the short dorsocentral hysterosomal setae taper distally. There are nine tactile setae and one solenidion on tibia I; there are five or six tactile setae on tibia II; there are five tactile setae on tibia III; and tarsus I has five tactile setae proximal to the duplex setae.

Collections have been made in Arizona on *Juniperus communis* L., *Thuja occidentalis* L., *Cupressus arizonica* Greene, and *Tamarix* sp.

106

Platytetranychus symphoricarpi, new series
(Figs. 108-112)

This species is related to *Platytetranychus libocedri,* having short dorsal setae and five tactile setae on tibia II. It differs in having rounded rather than sharp lobes on the striae and in having stronger leg setae.

Female. Terminal sensillum of palpus small, not two times as long as width; stylophore slightly indented anteriorly; peritreme bent distally and simple. First pair of propodosomal setae longer than other body setae which are subequal in length. Striae longitudinal on propodosoma and transverse on hysterosoma, with rounded lobes. Striae transverse on genital flap and on area anterior to flap. Leg setae strong, serrate. Tarsus I with a single solenidion; tibia I with nine strong tactile setae and one solenidion; genu I with five tactile setae; femur I with six tactile setae. Tarsus II with a single solenidion; tibia II with six tactile setae; genu II with five tactile setae; femur II with five tactile setae. Tarsus III with a slender solenidion; tibia III with five tactile setae; genu III with three tactile setae; femur III with three tactile setae. Tarsus IV with a slender solenidion; tibia IV with six tactile setae; genu IV with three tactile setae; femur IV with three tactile setae. Length of body 351 μ; including rostrum 446 μ.

Male. Not known.

Holotype. Female, *ex Symphoricarpos oreophilus* Gray, Portal, Arizona, August 28, 1964, by D. M. T.

Anatetranychus Womersley

Anatetranychus Womersley, 1940: 261; Pritchard and Baker, 1955: 215, 225; Wainstein, 1960: 200.

Type. Anatetranychus hakea Womersley.

Superficially this genus appears similar to *Oligonychus,* but differs in having two pairs of para-anal setae, and in that the empodial claw is bare, not possessing the ventrally directed proximoventral hairs.

The type was described from Western Australia. Two species have been found in Arizona.

KEY TO THE SPECIES OF ANATETRANYCHUS (FEMALES)

1. Dorsocentral setae as long as intervals between their bases 2
   Dorsocentral setae and dorsocentral setae not as long as distance between their bases . . . . *dalae,* new species
2. Dorsolateral and dorsocentral setae subequal in length
   . . . . *albiflorae,* new species
   Dorsolateral setae much longer than short dorsocentral setae . . . .
   . . . . *hakea* Womersley

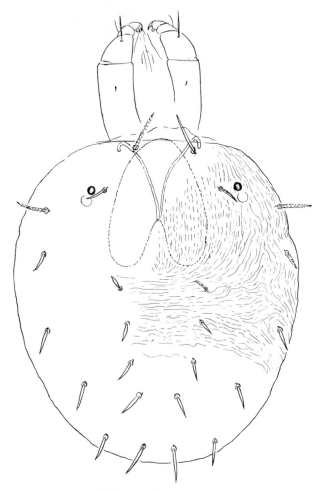

Figure 108, Dorsum of female.
Platytetranychus symphoricarpi, new species

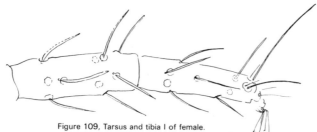

Figure 109, Tarsus and tibia I of female.
Platytetranychus symphoricarpi, new species

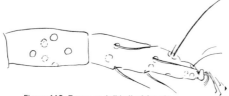

Figure 110, Tarsus and tibia II of female.
Platytetranychus symphoricarpi, new species

Figure 111, Tarsus and tibia III of female.
Platytetranychus symphoricarpi, new species

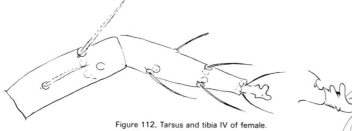

Figure 112, Tarsus and tibia IV of female.
Platytetranychus symphoricarpi, new species

109

Anatetranychus daleae, new species

(Figs. 113-118)

The species is distinctive in having the dorsal body setae shorter than the intervals between them. The aedeagus turns ventrally at a gentle angle, and has angulations distally.

Female. Terminal sensillum of palpus about twice as long as broad; rounded distally; stylophore broadly rounded; peritreme short and ending in a simple bulb. Striae longitudinal on propodosoma and transverse on hysterosoma, forming a V-pattern between the third pair of dorsocentral setae; striae with strong, rounded lobes. Dorsal body setae short, not as long as distance between their bases, strong and densely serrate. All empodia without proximoventral hairs; duplex setae of tarsus I distal and approximate; duplex setae of tarsus II shorter than on I. Tarsus I with three tactile setae and one solenidion proximal to duplex setae; tibia I with nine tactile setae and one solenidion; genu I with five tactile setae; femur I with nine tactile setae. Tarsus II with one solenidion; tibia II with six tactile setae; genu II with five tactile setae; femur II with seven tactile setae. Tarsus III with one dorsomedian solenidion; tibia III with six tactile setae; genu III with four tactile setae; femur III with four tactile setae. Tarsus IV similar to III; tibia IV similar to III; genu IV with three tactile setae; femur IV with three tactile setae. Length of body 357 μ; including rostroum 460 μ.

Male. Terminal sensillum of palpus small, slightly longer than broad; peritreme ending in a simple bulb; stylophore rounded anteriorly. Dorsal body setae short and serrate, as in female. All empodia slender and clawlike, without proximoventral hairs. Tarsus I with three solenidia; tibia I with ten tactile setae and three solendidia; genu I with five tactile setae; femur I with eight tactile setae. Tarsus II with one slender solenidion; tibia II with seven tactile setae; genu II with five tactile setae; femur II with seven tactile setae. Tarsus III with a dorsomedian solendidion; tibia III with six tactile setae; genu III with four tactile setae; femur III with four tactile setae. Tarsus IV with slender solendidion; tibia IV with six tactile setae; genu IV with three tactile setae; femur IV with three tactile setae. Aedeagus turns ventrad at a slight angle, and has angulation distally; the head of the aedeagus is at an angle with the shaft. Length of body 351 μ; including rostrum 434 μ.

Holotype. Female, *ex Dalea formosa* Torr., McNary, Arizona, June 27, 1962, by D. M. T.

Paratypes. Two males and seven females with the above data.

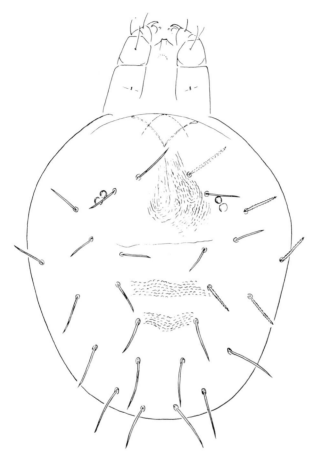

Figure 113, Dorsum of female.
Anatetranychus daleae, new species

111

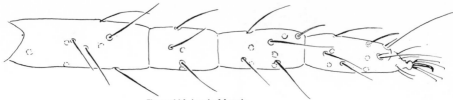

Figure 114, Leg I of female.
Anatetranychus daleae, new species

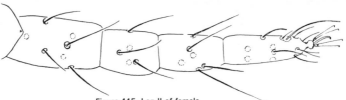

Figure 115, Leg II of female.
Anatetranychus daleae, new species

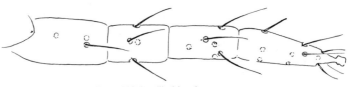

Figure 116, Leg III of female.
Anatetranychus daleae, new species

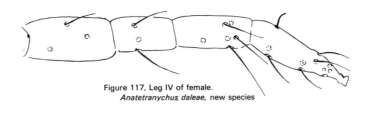

Figure 117, Leg IV of female.
Anatetranychus daleae, new species

Figure 118, Aedeagus.
Anatetranychus daleae, new species

112

Anatetranychus albiflorae, new species

(Figs. 119-122)

The long dorsal body setae and the aedeagus are distinctive of this species.

Male. Stylophore short, rounded anteriorly; peritreme ending in simple bulb. First pair of propodosomal setae about as long as distance between bases, shorter than other two pairs. Hysterosomal setae about as long as second and third pair of propodosomal setae, except for fifth pair of dorsocentral setae and fourth pair of dorsolateral setae which are shorter, the dorsocentral setae about two-thirds as long as the lateral setae. Tarsus I with two sets of distal duplex setae; four tactile setae and three solenidia proximal to duplex setae; tibia I with nine tactile setae and four solenidia; genu I with five tactile setae; femur I with nine tactile setae. Tarsus II with two tactile setae and one solenidia proximal to duplex setae; tibia II with six tactile setae; genu II with five tactile setae; femur II with seven tactile setae. Tarsus III with proximodorsal solenidion and ten tactile setae; tibia III with six tactile setae; genu III with four tactile setae; femur III with four tactile setae. Tarsus IV similar to III; tibia IV similar to III; genu IV with three tactile setae; femur IV with three tactile setae. Empodia without proximoventral hairs. Aedeagus bends abruptly distally, with angulations distally, the head of the aedeagus is parallel with the shaft. Length of body 300 μ; including rostrum 370 μ.

Female. Stylophore longer than wide and rounded anteriorly; peritreme ending in simple bulb. Dorsal body setae longer than intervals between their bases; first propodosomal and fifth dorsocentral hysterosomal setae shorter than others and subequal in length. Striae longitudinal on propodosoma and transverse on hysterosoma, with V-pattern between third pair of dorsocentral setae; lobes of striae low and rounded. Tarsus I with two sets of distal approximate duplex setae, and with four tactile setae and one solenidion proximal to duplex setae; tibia I with nine tactile setae and one slender distal solenidion; genu I with five tactile setae; femur I with nine tactile setae. Tarsus II with a single set of duplex setae and two tactile setae and one solenidion proximal to duplex setae (one tactile seta also lateral to duplex setae); tibia II with six tactile setae; genu II with five tactile setae; femur II with seven tactile setae. Tarsus III with slender dorsomedian solenidion and ten tactile setae; tibia III with six tactile setae; genu III with four tactile setae; femur III with four tactile setae. Tarsus IV similar to III; tibia IV similar to III; genu IV with three tactile setae; femur IV with three tactile setae. All empodial

113

claws strong, curved and without proximoventral hairs. Length of body 319 μ; including rostrum 434 μ.

Holotype. Male, *ex Berberis repens* Lindl., Prescott, Arizona, August 18, 1965, by D. M. T.

Paratypes. Three males and twelve females with the above data; one male and eighteen females *ex Dalea albiflora* Gray, Prescott, Arizona, August 19, 1965, by D. M. T.

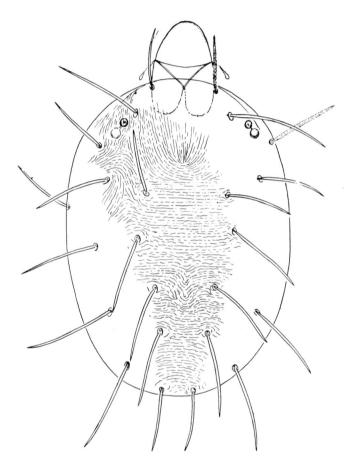

Figure 119, Dorsum of female.
Anatetranychus albiflorae, new species

114

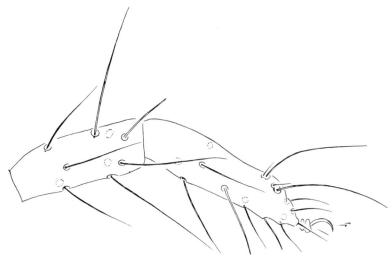

Figure 120, Tarsus and tibia I of female.
Anatetranychus albiflorae, new species

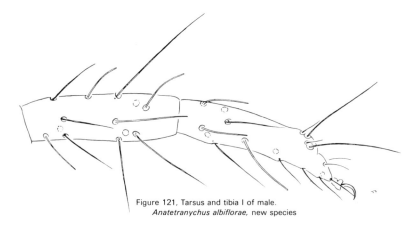

Figure 121, Tarsus and tibia I of male.
Anatetranychus albiflorae, new species

Figure 122, Aedeagus.
Anatetranychus albiflorae, new species

115

Tylonychus Miller

Tylonychus Miller, 1966: 59.

Type. Tylonychus tasmaniensis Miller.

This genus is similar to *Anatetranychus* with the absence of proximo-ventral hairs on the clawlike empodium. It differs from *Anatetranychus* by the presence of spinules on the striae, stout setae on tubercles, and the unusually long gnathosoma.

The genus is represented by a single species (*Tylonychus tasmaniensis* Miller) from Tasmania occurring on *Acacia mearnsii* de Wild and *Acacia dealbata* Link.

Mixonychus Ryke and Meyer

Myxonychus[1] Ryke and Meyer, 1960: 559.

Type. Myxonychus[1] *acaciae* Ryke and Meyer.

The peritremes are simple and slightly bent distally; the stylophore is indented anteriorly. The dorsal body setae are normal in number and position for the Tetranychini. The duplex setae of tarsus I are distal and approximate; the empodial claw is without proximoventral hairs. The dorsal integument is reticulate.

The genus is known only from South Africa. The only known species is from *Acacia karoo* Hayne.

Dr. Magdelena Meyer has written (December 1965) that she has found another species belonging to this genus.

Oligonychus Berlese

Oligonychus Berlese, 1886: 24; Pritchard and Baker, 1955: 270;
    Wainstein, 1960: 203.

Type. Heteronychus brevipodus Targioni Tozzetti.

Mites of the genus *Oligonychus* possess a single pair of para-anal setae; the empodium is well developed and clawlike with proximoventral hairs set at right angles to the claw; and the dorsal body setae, with few exceptions, are not set on tubercles.

The type, an immature, was taken from holly oak. The setae are figured as being long, probably as in *Oligonychus bicolor* (Banks). Wainstein (1960) illustrates a mite with short dorsal setae as *O. brevipodus*. Mites with these short setae, are, as far as we know, conifer feeders and are not found on oak. Consequently we have a somewhat different conception of *Oligonychus* s. str. from that of Wainstein. We also disagree with the placement of some species into subgenera.

---

[1] *Mixonychus* is the correct spelling. The "y" was in error in the original description of the genus, but "i" was used throughout in the abstract and in the specific description, etc.

Wainstein (1960) divided *Oligonychus* into five subgenera, which he further subdivided. His subgeneric divisions are:

Oligonychus Berlese. *Type. Heteronychus brevipodus* Targioni Tozzetti.
Homonychus Wainstein. *Type. Tetranychus peruvianus* McGregor.
Pritchardinychus Wainstein. *Type. Paratetranychus pritchardi* McGregor.
Metatetranychoides Wainstein. *Type. Paratetranychus (Metatetranychoides) quercifolius* Wainstein.
Paratetranychus Zacher. *Type. Tetranychus ununguis* Zacher.

We have divided the genus into six subgenera as follows:

Oligonychus Berlese. Type. Heteronychus brevipodus Targioni Tozzetti.
Wainsteiniella, new subgenus. *Type. Paratetranychus subnudus* McGregor.
Homonychus Wainstein. *Type. Tetranychus peruvianus* McGregor.
Metatetranychoides Wainstein. *Type. Paratetranychus (Metatetranychoides) quercifolius* Wainstein.
Reckiella, new subgenus. *Type. Tetranychus pratensis* Banks.
Pritchardinychus Wainstein. *Type. Paratetranychus pritchardi* McGregor.
We consider *Paratetranychus* Zacher a synonym of *Oligonychus* s. str.

KEY TO THE SPECIES OF OLIGONYCHUS OF ARIZONA (FEMALES)

1. Striae of hysterosoma transverse 2
   Striae of hyterosoma longitudinal between third or fourth pair of dorso-central setae 5
2. Hysterosomal setae much shorter than distance between their bases
   . . . . (*Wainsteiniella*) 3
   Hysterosomal setae as long as distance between their bases 4
3. Tarsus II with solenidion . . . . *subnudus* (McGregor)
   Tarsus II without solenidion . . . . *milleri* (McGregor)
4. Tibia I with seven tactile setae (*Oligonychus* s. str.) . . . . . .
   . . . . *ununguis* (Jacobi) and *coniferarum* (McGregor)
   Tibia I with nine tactile setae . . . . (*Pritchardinychus*) *pritchardi*
   . . . . (McGregor) and *propetes* Pritchard and Baker
5. Striae longitudinal between third pair of dorsocentral setae . . . .
   . . . . (*Homonychus*) 6
   Striae longitudinal between fourth pair of dorsocentral setae *(Reckiella)*
   . . . . . *pratensis* (Banks), *modestus* (Banks), *stickneyi* (McGregor),
   and *keiferi,* new species
6. Duplex setae of tarsi I and II with units nearly equal in length . . .
   . . . . *platani* (McGregor)
   Duplex setae with units greatly differing in length, *gambelii,* new species

Where several species are listed in the same couplet a profile of the male aedeagus is necessary to separate these mites.

117

Oligonychus (*Oligonychus*) **Berlese**

Type. Heteronychus brevipodus Targioni Tozzetti.

The hysterosomal striae are transverse; the dorsal setae are longer than the distances between their bases; and the aedeagus is downturned at an obtuse angle. They generally feed upon the upper surfaces of broadleaf plants, although a few feed on conifers.

The identity of *Oligonychus brevipodus* is unknown, but since it was collected from holly oak, and because of the long body setae as originally figured, it is probably related to *O. bicolor* (Banks), etc. At present, *O. ununguis* (Jacobi) and *O. coniferarum* (McGregor) are the only two species known from Arizona. Both are conifer feeders.

Oligonychus (*Oligonychus*) *ununguis* (Jacobi)

Tetranychus ununguis Jacobi, 1905: 239.
Oligonychus ununguis, Tuttle and Baker, 1964: 34.

The dorsal body setae are much longer than the distance between their bases. The hysterosomal striae are transverse. There are four tactile setae and usually one solenidion proximal to the duplex setae on tarsus I; there are seven tactile setae and usually one solenidion on tibia I. The aedeagus is distinctive in that the bend is at right angles to the shaft, and the bent portion tapers gradually to an acute tip, the ventrally directed portion being about one-half as long as the dorsal margin of the shaft.

This species is a pest of conifers throughout the world. It has been taken in Arizona on *Abies lasiocarpa* (Hook.) Nutt., *Cupressus arizonica* Greene, *Cupressus sempervirens* L., *Juniperus communis* L., *Juniperus deppeana* Steud., *Picea engelmannii* Parry, *Picea pungens* Engelm., *Pinus ponderosa* Lawson, *Pseudotsuga menziesii* (Mirb.) Franco, and *Thuja occidentalis* L.

Oligonychus (*Oligonychus*) *coniferarum* (McGregor)

Paratetranychus coniferarum McGregor, 1950: 338.
Oligonychus coniferarum, Tuttle and Baker, 1964: 36.

This species is similar to *Oligonychus ununguis,* differing only in that the aedeagus forms a short, truncate, caudolaterally directed bend. The angle is obtuse, and the bent portion is very short.

Oligonychus coniferarum has been collected in Arizona on *Cupressus sempervirens* L. and *Thuja occidentalis* L.

Oligonychus (*Wainsteiniella*), **new subgenus**

Type. Paratetranychus subnudus McGregor.

The dorsal setae are shorter than the intervals between their bases; the hysterosomal striae are transverse; and the aedeagus is downturned.

These mites are conifer feeding. *Oligonychus milleri* (McGregor), *O. subnudus* (McGregor), *O. cunliffei* Pritchard and Baker, and *O. pityinus* Pritchard and Baker belong here. The first two species have been found in Arizona.

Oligonychus (*Wainsteiniella*) *subnudus* (McGregor)

Paratetranychus subnudus McGregor, 1950: 355.
Oligonychus subnudus, Tuttle and Baker, 1964: 36.

The dorsocentral setae are short and subequal in length. The tarsi are slender and longer than the tibia, especially on legs III and IV; there is a single seta proximal to the duplex setae on tarsus I; there is a long lateral solenidion on tarsus II. The stylophore is indentate anteriorly. The aedeagus is bent ventrally at right angles but is not as long nor as acutely angled as in the related species *Oligonychus milleri* (McGregor).

Oligonychus subnudus has been taken in Arizona on *Pinus ponderosa* Lawson, *P. edulis,* Engelm., *Asclepias speciosa* Torr., and *Antennaria arida* E. Nels.

Oligonychus (*Wainsteiniella*) *milleri* (McGregor)

Paratetranychus milleri McGregor, 1950: 343.
Oligonychus milleri, Tuttle and Baker, 1964: 36.

This species has the dorsocentral setae successively increasing in length, the first pair being very short, the second pair longer, and the third pair much longer than the first. There is a single tactile setae proximal to the duplex setae on tarsus I; there is no solenidion on tarsus II.

Oligonychus milleri has been collected in Arizona on *Antennaria arida* E. Nels., *Asclepias speciosa* Torr., *Euonymus* sp., *Olea europaea* L., *Photinia arbutifolia* (Ait.) Lindl., and *Pinus* sp.

Oligonychus (*Homonychus*) Wainstein

Oligonychus (*Homonychus*) Wainstein, 1960: 216.

Type. Tetranychus peruvianus McGregor.

The striae of the hysterosoma are transverse except for the longitudinal pattern between the third pair of dorsocentral setae. The aedeagus is downturned.

Oligonychus platani (McGregor) and *O. gambelii,* new species, which are found in Arizona, are included here, although the setal pattern of the legs vary somewhat from that of the type species. *O. peruvianus* has nine tactile setae on tibia I, *O. platani* seven, and *O. gambelii* six.

Oligonychus (*Homonychus*) *platani* (McGregor)

Paratetranychus platani McGregor, 1950: 349.
Oligonychus platani, Tuttle and Baker, 1964: 36.

This species is unique in that the members of the duplex setae are short and more or less equal in length. There are seven tactile setae on tibia I; and three or four tactile setae proximal to the duplex setae on the tarsus I of the female. The striae of the female are either longitudinal or with a V-shaped pattern between the third pair of dorsocentral setae. The aedeagus is distinctive in that the distal fourth bends abruptly ventrad, the distal end being slender.

Oligonychus platani has been collected in Arizona on *Eriobotrya japonica* (Thunb.) Lindl., *Pyracantha coccinea* Roem., and *Quercus emoryi* Torr.

Oligonychus (*Homonychus*) *gambelii*, new species

(Fig. 123)

The striae of the female are longitudinal between the third pair of dorsocentral setae of the hysterosoma, and the units of the duplex setae are unequal in length.

Female. Rostrum short and broad; stylophore broad and rounded anteriorly with slight indentation; peritreme ending in a simple bulb. Dorsal striae with strong, rounded lobes; striae longitudinal on propodosoma and transverse on hysterosoma except for area between third pair of dorsocentral setae. Dorsal setae long, strong, and strongly serrate; anterior pair of prodosomal setae about one-half as long as second pair and about same length as third pair; fifth pair of dorsocentral setae small and slender; fourth pair of dorsolateral setae and humeral setae stronger and longer, about one-half as long as dorsocentral setae. Tibia I with six setae and one solenidion; tarsus I with two tactile setae and one solenidion proximal to the duplex setae. Tibia II with five setae. Tibia III and IV each with three setae; tarsi III and IV each with a proximal solenidion and eight tactile setae. Length of body 268 μ; including rostrum 319 μ.

Male. Not known.

Holotype. Female, *ex Quercus gambelii* Nutt., McNary, Arizona, July 24, 1964, by D. M. T.

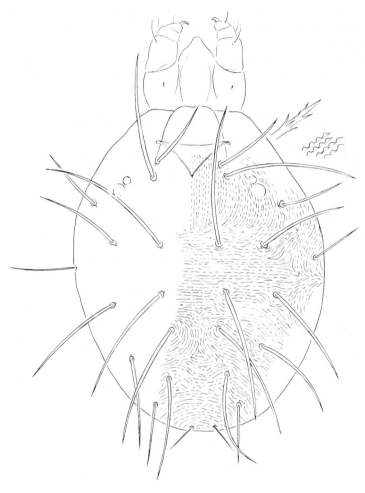

Figure 123, Dorsum of female.
Oligonychus (Homonychus) gambelii, new species

Oligonychus (*Metatetranychoides*) Wainstein

Oligonychus (Metatetranychoides) quercifolius Wainstein.

Type. Paratetranychus (Metatetranychoides) quercifolius Wainstein.

These mites are darkish in color; the hysterosomal striae are transverse except for an irregular pattern between the third pair of dorsocentral setae. The aedeagus is downturned. They feed on the lower surface of leaves of broad-leaf plants.

Two American species, *Oligonychus endytus* Pritchard and Baker, and *O. aceris* (Shimer), belong to this subgenus. These species have not yet been found in Arizona.

Oligonychus (*Reckiella*), new subgenus

Type. Tetranychus pratensis Banks.

The striae are transverse on the hysterosoma except for the longitudinal pattern between the fourth pair of dorsocentral setae. The aedeagus is upturned, and has an anterior and posterior hook or angulation. In most cases these mites feed on monocot plants.

Oligonychus pratensis (Banks), *O. modestus* (Banks), *O. stickneyi* (McGregor), and *O. keiferi,* new species have been found in Arizona.

Oligonychus (*Reckiella*) *pratensis* (Banks)

Tetranychus pratensis Banks, 1912: 97.

Oligonychus pratensis, Tuttle and Baker, 1964: 38.

The dorsal striae of the female are longitudinal between the fourth pair of dorsocentral setae. The peritreme ends in a simple bulb. There are nine tactile setae on tibia I of the female. The aedeagus is upturned and the knob is about twice as wide as the stem of the knob.

This species has been found on many grasses, both wild and cultivated, throughout Arizona by D. M. T. including: *Agropyron repens* (L.) Beauv., *Agropyron trachycaulum* (Link) Malte, *Agrostis alba* L., *Aristida adscensionis* L., *Arundiraria simoni* (Carr.), *Bouteloua barbata* Lag., *Bouteloua rothrockii* Vasey, *Bromus anomalus* Rupr., *Bromus arizonicus* (Sheer) Stebbins, *Carex simulata* Mackenz., *Cenchrus echinatus* L., *Chloris virgata* Swartz, *Cucurbita palmata* Wats., *Cyperus alternifolius* L., *Cynodon dactylon* (L.) Pers., *Echinochloa colonum* (L.) Link, *Elymus canadensis* L., *Elymus pauciflorus pseudorepens* Gould, *Eriochloa lemnoni* Vasey & Scribn., *Euphorbia albomarginata* Torr. & Gray, *Hilaria rigida* (Thurb.) Benth., *Hordeum jubatum* L., *Leptochloa uninervia* (Presl.) Hitchc. & Chase, *Lolium perenne* L., *Panicum antidotale* Retz., *Panicum capillare* L., *Paspalum dilatatum* Poir., *Petunia parviflora* Juss., *Phalaris minor* Retz., *Phragmites communis* Trin., *Poa pratensis* L., *Sitanion hystrix* (Nutt.) J. G. Smith, *Setaria viridis* (L.) Beauv., *Sorghum halepense* (L.) Pers., *Sorghum vulgare* Pers., *Tribulus terrestris* L., and *Zea mays* L.

Oligonychus (*Reckiella*) *modestus* (Banks)

Tetranychus modestus Banks, 1900: 73.
Oligonychus modestus, Pritchard and Baker, 1955: 355; Tuttle and Baker, 1964: 38.

This species is similar to *Oligonychus pratensis.* It differs in that the knob of the aedeagus is not two times as long as the stem of the knob is wide, and that the axis of the knob is parallel to the axis of the shaft.

Oligonychus modestus has been reported on *Zea mays* L., in Arizona.

Oligonychus (*Reckiella*) *stickneyi* (McGregor)

Paratetranychus stickneyi McGregor, 1939: 253.
Oligonychus stickneyi, Tuttle and Baker, 1964: 38.

This species is similar to *Oligonychus pratensis,* differing only in that the aedeagal knob is much larger, being about one-third as long as the dorsal portion of the shaft, the axis of the knob forming less than a 45° angle with the axis of the shaft.

Oligonychus stickneyi has been taken from various species of Gramineae in Arizona including: *Bouteloua barbata* Lag., *Bouteloua curtipendula* (Michx.) Torr., *Chloris virgata* Swartz, *Cynodon dactylon* (L.) Pers., *Echinochloa crusgalli* (L.) Beauv., *Eriochloa lemmoni* Vasey and Scribn., *Lolium perenne* L., *Paspalum dilatatum* Poir., *Sorghum vulgare* Pers., and *Zea saccharata* Sturtev.

Oligonychus (*Reckiella*) *keiferi*, new species

(Fig. 124)

The shape of the aedeagus is distinctive.

Female. Typical for the subgenus in that the striae are transverse between the fourth pair of dorsocentral hysterosomal setae. It is similar to all other members of the subgenus. Length of body 319μ; including rostrum 414 μ.

Male. Typical for the subgenus except in the shape of the aedeagus which is somewhat similar to that of *Oligonychus stickneyi* (McGregor) but being stronger and more rounded anteriorly. Length of body 240 μ; including rostrum 320 μ.

Holotype. Male, *ex Lycurus phleoides* H.B.K., Prescott, Arizona, August 19, 1965, by D. M. T.

Paratypes. Two males and nine females with the same data as above.

Figure 124, Aedeagus.
Oligonychus (Reckiella) keiferi, new species

123

Oligonychus (*Pritchardinychus*) Wainstein

Oligonychus (*Pritchardinychus*) Wainstein, 1960: 217.

Type. Paratetranychus pritchardi (McGregor).

The hysterosomal striae of the female are transverse. The aedeagus is upturned and then ventrally directed with both anterior and posterior angulations. There are nine tactile setae on tibia I. These are small green mites feeding on the ventral leaf surfaces.

We consider only *Oligonychus pritchardi* (McGregor), *O. propetes* Pritchard and Baker, and *O. hadrus* Pritchard and Baker to belong to this subgenus.

Oligonychus (*Pritchardinychus*) *pritchardi* (McGregor)

Paratetranychus pritchardi McGregor, 1950: 350.
Oligonychus pritchardi, Pritchard and Baker, 1955: 365.

The female resembles that of *Oligonychus propetes* Pritchard and Baker. The aedeagus is distinctive in that the distal hook is nearly straight on the dorsal margin, except for the obtuse angulation.

This species has been collected on *Quercus gambelii* Nutt., McNary, Arizona, February 20, 1965, by D. M. T.

Oligonychus (*Pritchardinychus*) *propetes* Pritchard and Baker

Oligonychus propetes Pritchard and Baker, 1955: 366.

The female is small, greenish, and feeds on the undersurface of leaves of broad-leaf plants. The peritreme is straight distally. There are four tactile setae and one solenidion proximal to the duplex setae on tarsus I; there are nine tactile setae and one solenidion on tibia I. The striae of the female are transverse on the hysterosoma and longitudinal on the propodosoma. The aedeagus has a distal enlargement strongly curved on the dorsal side and bent ventrally.

Oligonychus propetes has been collected on *Quercus arizonica* Sarg., Portal, Arizona, August 28, 1964, by D. M. T.

Tetranychus Dufour

Tetranychus Dufour, 1832: 276; Pritchard and Baker, 1955: 373;
Wainstein, 1960: 149.

Type. Tetranychus lintearius Dufour.

There is a single pair of para-anal setae. The empodium has free proximoventral hairs, and, if a clawlike dorsal member is present, it is much shorter than the hairs; the duplex setae of tarsus I are widely separate, dividing the segment into three more or less equal parts. The empodium of tarsus I of the male usually consists of short tridigitate appendages with or without a medial spur. The aedeagus always bends dorsally. Setae are not borne on tubercles in either sex.

124

Adult females in the northern regions are usually greenish and overwinter as nonfeeding orange-colored females. Those in the more tropical areas may reproduce throughout the year, and are usually reddish in color. Wainstein (1960) divided the genus into seven subgenera as follows:

Tetranychus Dufour. *Type. Tetranychus lintearius* Dufour.

Polynychus Wainstein. *Type. Tetranychus homorus* Pritchard and Baker.

Armenychus Wainstein. *Type. Tetranychus armeniaca* Bagdasarian.

Pentanychus Wainstein. *Type. Tetranychus fijiensis* Hirst.

Septanychus McGregor. *Type. Tetranychus tumidus* Banks.

Pseudonychus Wainstein. *Type. Tetranychus ludeni* Zacher.

Amphitetranychus Oudemans. *Type. Tetranychus viennensis* Zacher.

In general, these groups are based upon the arrangement of the empodial rays, the peritremes, and the color of the feeding females.

In the classification we are submitting, we are relying entirely upon the striation pattern of the females for the subgeneric groupings. We propose the following, with the types as above:

Tetranychus Dufour.

Polynychus Wainstein.

Armenychus Wainstein.

Tetranychus s. str. includes the mites with the dorsal striae of the female forming a diamond-shaped pattern between the third and fourth pairs of dorsocentral setae. We believe this subgenus includes *Pentanychus, Septanychus, Pseudonychus,* and *Amphitetranychus.* It is possible that each of the above may form groups within the subgenus, but this is beyond the scope of the paper.

KEY TO THE SPECIES OF TETRANYCHUS IN ARIZONA (FEMALES)

1. Striae of hysterosoma not transverse in entirety 2
   Striae transverse on hysterosoma (*Armenychus*) . . . .
   . . . . *pacificus* McGregor, *mcdanieli* McGregor

2. Striae longitudinal between third and fourth pair of dorsocentral setae forming a diamond-shaped figure between these setae (*Tetranychus* s.str.) 3
   Striae longitudinal between fourth pair of dorsocentral setae (*Polynchus*)
   . . . . . . *canadensis* (McGregor), *polys* Pritchard and Baker

3. Proximal pair of duplex setae anterior to the four proximal tactile setae of tarsus I 4
   Proximal pair of duplex setae in line with the four proximal tactile setae of tarsus I . . . . *desertorum*
   . . . . . Banks, *gigas* Pritchard and Baker, *yucca,* new species.

4. Female greenish; lobes of striae rounded . . . .
   . . . . *atlanticus* McGregor and *uritcae* Koch
   Female red; lobes of striae pointed and as tall as broad . . . . .
   . . . . *cinnabarinus* (Boisduval)

125

Tetranychus (*Tetranychus*) s. str.

Tetranychus Dufour, 1832: 276; Wainstein, 1960: 149.

Type. Tetranychus lintearius Dufour.

The striae form a diamond pattern between the third and fourth pairs of dorsocentral setae. Most of the species in the genus belong here. In Arizona the subgenus *Tetranychus s. str.* contains *T. desertorum, T. gigas, T. yucca, T. atlanticus, T. cinnabarinus,* and *T. urticae.*

Tetranychus (*Tetranychus*) *desertorum* Banks

Tetranychus desertorum Banks, 1900: 76; Pritchard and Baker, 1955: 403; Tuttle and Baker, 1964: 40.

The hysterosomal striae are longitudinal between the third and fourth pairs of dorsocentral setae and form a diamond pattern between these setae. The peritreme is hooked distally. The proximal pair of duplex setae of tarsus I are in line with the four proximal tactile setae; minute empodial spurs may be present. The dorsal margin of the knob of the aedeagus is sigmoid, the anterior angulation is small and acute and the posterior angulation is acute and curved ventrad to a variable extent; the width of the knob is about one-fourth as long as the dorsal margin of the shaft.

This species has been taken on many plants throughout Arizona by D. M. T. as follows: *Ambrosia ambrosiodes* (Cav.) Payne, *Ambrosia dumosa* (Gray) Payne, *Apium graveolens* L., *Asclepias erosa* Torr., *Aster spinosus* Benth., *Astragalus arizonicus* Gray, *Atriplex canescens* (Pursh) Nutt., *Atriplex confertifolia* (Torr. & Frém.) Wats., *Atriplex elegans* (Moq.) D. Dietr., *Baileya multiradiata* Harv. & Gray, *Baileya pleniradiata* Harv. & Gray, *Beta vulgaris* L., *Bouteloua curtipendula* (Michx.) Torr., *Brassica nigra* (L.) Koch, *Citrullus vulgaris* Schrad., *Coldenia palmeri* Gray, *Cressa truxillensis* H.B.K., *Cucumis melo* L., *Cucurbita digitata* Gray, *Cucurbita palmata* Wats., *Dalea mollis* Benth., *Datura stramonium* L., *Ditaxis serrata* (Torr.) Heller, *Echinocereus dasyocanthus* Engelm., *Encelia frutescens* Gray, *Erodium cicutarium* (L.) L'Hér, *Eriogonum abertianum* Torr., *Fragaria chiloensis* Duchesne, *Gardenia* sp., *Gaura coccinea* Nutt., *Glycine max* (L.) Merr., *Gossypium hirsutum* L., *Haplopappus acradenius* (Green) Blake, *Haplopappus spinulosus* (Pursh) DC., *Hymenoclea pentalepis* Rydb., *Hymenothrix wislizeni* Gray, *Lactuca sativa* L., *Lantana camara* L., *Larrea tridentata* (DC.) Colville, *Lathryus odoratus* L., *Lepidium thurberi* Wooton, *Lupinus* sp., *Malva parviflora* L., *Marrubium vulgare* L., *Medicago sativa* L., *Mentzelia pumila* (Nutt.) Torr. & Gray, *Nama hispidum* Gray, *Oxybaphus comatus* (Small) Weatherby, *Perezia nana* Gray, *Simmondsia chinensis* (Link) Schneid., *Sorghum vulgare* Pers., *Sphaeralcea* sp., *Stephanomeria pauciflora* (Torr.) A. Nels., *Suaeda torreyana* Wats., *Tribulus terrestris* L., *Verbena bipinnatifida* Nutt., *Xanthium saccharatum* Wallr., *Zea saccharata* Sturtev., and *Zinnia elegans* Jacq.

Tetranychus (*Tetranychus*) *gigas* Pritchard and Baker

Tetranychus gigas Pritchard and Baker, 1955: 405.

This species is similar to *Tetranychus desertorum* Banks, differing in that the distal knob of the aedeagus is large, being about one-third as long as the dorsal margin of the shaft.

Tetranychus gigas has been collected on cotton (*Gossypium hirsutum* L.) in Texas and Arizona.

Tetranychus (*Tetranychus*) *yuccae*, new species

(Fig. 125)

Tetranychus yuccae is similar to *T. desertorum* Banks in having the duplex and proximal tactile setae of tarsus I of the female in line. The aedeagus is distinctive in being sigmoid with a slight anterior angulation, an indented dorsal margin, and a strong almost straight posterior angulation.

Male. Terminal sensillum of palpus about two times as long as broad; stylophore broad and rounded anteriorly; peritreme hooked distally. Dorsal setae much longer than distances between bases. Tarsus I with strong empodial spur and with other elements fused; tarsus I with three long slender solenidia; tibia I with four long slender solenidia and nine tactile setae; genu I with five tactile setae; femur I with ten tactile setae. Empodium of tarsus II with dorsal spur and clawlike hairs; tarsus II with one long slender proximal solenidion; tibia II with seven tactile setae; genu II with five tactile setae; femur II with six tactile setae. Tarsus III with small dorsal empodial spur and free empodial hairs and with one long proximal solenidion; tibia III with six tactile setae; genu III with four tactile setae; femur III with four tactile setae. Leg IV similar to leg III except tibia IV with seven tactile setae. Aedeagus short, stout, as figured; head at slight angle to shaft, indented dorsally and with slight angulation anteriorly and with strong slightly downcurved angulation posteriorly. Length of body including rostrum 573 μ.

Female. Terminal sensillum of palpus not more than two times as long as broad; stylophore short, broad and rounded anteriorly; peritreme hooked distally. All dorsal setae much longer than distance between their bases. Striae forming a diamond-shaped pattern between the third and fourth pairs of dorsocentral hysterosomal setae; lobes of striae as tall as broad. Proximal

Figure 125, Aedeagus.
Tetranychus (Tetranychus) yuccae, new species

127

tactile setae of tarsus I in line with proximal duplex setae; tibia I with one long slender solenidion and nine tactile setae; genu I with five tactile setae; femur I with ten tactile setae. Tarsus II with a long slender proximal solenidion; tibia II with seven tactile setae; genu II with five tactile setae; femur II with six tactile setae. Tarsus III with long slender proximal solenidion; tibia III with six tactile setae; genu III with four tactile setae. Femur III with four tactile setae. Leg IV similar to leg III except for tibia with seven tactile setae. All empodia with small dorsal spurs. Length of body 546 μ; including rostrum 653 μ.

Holotype. Male, *ex Yucca aloifolia* L., Tucson, Arizona, October 2, 1963, by Leon Moore.

Paratypes. Six males and three females with the above data.

Tetranychus (Tetranychus) atlanticus McGregor

Tetranychus atlanticus McGregor, 1941: 43: 26; Pritchard and Baker, 1955: 424.

The adult females are greenish or straw-colored, with the lobes of the striae rounded. The male is distinctive in that the distal knob of the aedeagus is moderately enlarged, being about one-fourth as long as the dorsal margin of the shaft, and the dorsal margin of the knob is obtusely angulate. The anterior projection of the knob is strong and narrowly rounded, and the posterior angulation is small and acute. The axis of the knob forms a slight angle with the axis of the shaft.

Tetranychus atlanticus has been taken in Arizona on *Ambrosia confertiflora* (DC.) Rydb., *Ambrosia dumosa* Gray, *Asclepias subulata* Decne., *Calendula officinalis* L., *Carthamus tinctorious* L., *Ceanothus fendleri* Gray, *Citrus sinensis* (L.) Osbeck, *Convolvulus arvensis* L., *Cucumis melo* L. *Cucumis sativus* L., *Daucus carota* L., *Dicoria canescens* Gray, *Euonymus europaeus* L., *Glycine max* (L.) Merr., *Gossypium hirsutum* L., *Heterotheca subaxillaris* (Lam.) Britt. and Rusby, *Lactuca sativa* L., *Medicago sativa* L., *Morsus rubra* L., *Polygonum argyrocoleon* Steud., *Raphanus sativus* L., *Ricinus communis* L., *Tribulus terrestris* L., *Trifolium alexandrinum* L., *Solanum elaeagnifolium* Cav., *Thalictrum fendleri* Engelm., *Viola conspersa* Reichens, *Zea mays* L., and *Zea mays saccharata* Sturtev.

Dosse and Boudreaux (1963) state that Wainstein (1961) has synonymized this species under *Tetranychus turkestani* Ugarov and Nikolski, 1937. We have not seen Wainstein's paper, but the wide distribution of *T. atlanticus* indicates that this synonymy may be correct.

Tetranychus (*Tetranychus*) *urticae* Koch

Tetranychus urticae Koch, 1836: 10; Boudreaux and Dosse, 1963: 363.
Tetranychus telarius (Linn.) of various authors.

Tetranychus urticae has only recently been used for the common green two-spotted spider mite of the temperate regions. Previously, *T. bimaculatus* Harvey and *T. telarius* (Linn.) were in common usage. Boudreaux and Dosse (1963) have shown that the proper name for the green species should be *T. urticae* Koch, and we concur.

This green species is well known in deciduous fruit trees in the northern regions of the United States and Europe and has been collected in the higher and cooler altitudes of Arizona on *Linaria dalmatica* Mill., and *Potentilla norvegica* L.

Tetranychus (*Tetranychus*) *cinnabarinus* (Boisduval)

Acarus telarius Linnaeus, 1758: 616; Boudreaux and Dosse, 1963: 363.
Acarus cinnabarinus Boisduval, 1867: 88.
Tetranychus cinnabarinus (Boisduval), Boudreaux, 1956: 43.

Various names have been proposed for this common red tetranychid found throughout the warmer parts of the world. By agreement at the 2nd International Congress of Acarology the name *T. cinnabarinus* will be submitted to the International Commission on Zoological Nomenclature for approval.

The female is red, with the lobes of the striae taller than broad. The aedeagus has an extremely small and nearly evenly shaped distal knob; some variation in the shape of the anterior angle of the knob may occur.

T. cinnabarinus has been collected on many hosts throughout the warmer parts of the world. It has been collected in Arizona on *Althaea Rosea* (L.) Cuv., *Apium graveolens* L., *Arachis hypogaea* L., *Aralia* sp. *Arctium minus* Schkuhr., *Artemisia dracunculoides* Pursh, *Aucuba japonica* Thunb., *Aster* sp., *Atriplex canescens* (Pursh) Nutt., *Atriplex lentiformis* (Torr.) Wats., *Atriplex semibaccata* R. Br., *Beta vulgaris* L., *Bouteloua barbata* Lag., *Brassica oleracea* L., *Calendula officinalis* L., *Capsicum frutescens* L., *Carthamus tinctorius* L., *Centaurea cyanus* L., *Centaurea imperialis* Hausskn, *Chaenactis stevioides* Hook and Arn., *Chenopodium murale* L., *Chrysanthemum* sp., *Citrullus vulgaris* Schrad., *Citrus limon* Burm. F., *Citrus* spp., *Citrus sinensis* (L.) Osbeck, *Convolvulus arvensis* L., *Cucumis melo* L., *Cucumis sativus* L., *Cucurbita pepo* L., *Cynodon dactylon* (L.) Pers., *Dalea mollis* Benth., *Ficus carica* L., *Fragaria chiloensis* Duchesne, *Galium stellatum*

Kellogg, *Gladiolus hortulanus* Bailey, *Glycine max* (L.) Merr., *Gossypium hirsutum* L., *Hedera helix* L., *Helianthella quinquenervis* (Hook.) Gray, *Hibiscus mutabilis* L., *Ipomoea* sp., *Jasminum* sp., *Lactuca lonifolia* Lam., *Lactuca sativa* L., *Lathyrus odoratus* L., *Lonicera* sp., *Lupinus argenteus* Pursh, *Lycopersicum esculentum* Mill., *Malus sylvestris* Mill., *Malva parviflora* L., *Medicago sativa* L., *Melia azedarach* L., *Morus rubra* L., *Musa nana* Lour., *Nama hispidum* Gray, *Oenothera clavaeformis* Torr. and Frém., *Nicotiana* sp., *Phaseolus acutifolius* A. Gray, *Philodendron* sp., *Physalis wrightii* Gray, *Pisum sativum* L., *Polygonum argyrocoleon* Steud., *Polygonum aviculare* L., *Populus* sp., *Pyracantha* spp., *Pyrus communis* L., *Raphanus sativus* L., *Ricinus communis* L., *Rosa centifolia* L., *Rosa dilecta* Rehd., *Rosa multiflora* Thunb., *Rosa* sp., *Sapindus* sp., *Sida hederacea* (Dougl.) Torr., *Sitanion hystrix* (Nutt.) J. G. Smith, *Sonchus asper* (L.) Hill, *Tragopogon dubius* Scop., *Tagetes erecta* L., *Thalictrum fendleri* Engelm., *Tribulus terrestris* L., *Verbena bracteata* Lag. & Rodr., *Vicia pulchella* H.B.K., *Vigna sinensis* (Torner) Savi, *Viola* sp., and *Zea mays saccharata* Sturtev.

Tetranychus (*Polynychus*) Wainstein

Tetranychus (*Polynychus*) Wainstein, 1960: 149.

Type. *Tetranychus homorus* Pritchard and Baker.

The striae are longitudinal between the fourth pair of dorsocentral setae. Two species of the subgenus *Polynychus* occur in Arizona, *Tetranychus canadensis* and *T. polys.*

Tetranychus (*Polynychus*) *canadensis* (McGregor)

Septanychus canadensis McGregor, 1950: 319.
Tetranychus canadensis, Tuttle and Baker, 1964: 41.

The hysterosomal striae of the female are longitudinal between the fourth pair of dorsocentral setae. The proximal pair of duplex setae are distad of proximal tactile setae. The peritreme is hooked distally. There are no empodial spurs. Empodium I of the male consists of short, tridentate appendages with a dorsomedial spur; empodium II is normal but has a strong dorsomedial spur; Empodia III and IV each have a small spur. The aedeagal knob is about one-third as long as the dorsal margin of the shaft; the anterior projection of the knob is rounded, and the caudal projection forms an acute angle. There is some variation in the size of the knob. The female is greenish with darker shoulder and posterior spots; the overwintering females are bright orange.

This species is common in Arizona and has been collected from *Atriplex canescens* (Pursh) Nutt., *Atriplex semibaccata* R. Br., *Atriplex polycarpa* (Torr.) Wats., *Cryptantha angustifolia* (Torr.) Greene, *Echinochloa colonum* (L.) Link, *Fraxinus velutina* Torr., *Melia azedarach* L., *Potentilla hippiana* Lehm., and *Thalictrum fendleri* Engelm.

130

Tetranychus (*Polynychus*) *polys* Pritchard and Baker

Tetranychus polys Pritchard and Baker, 1955: 396.

The striae are longitudinal between the fourth pair of dorsocentral hysterosomal setae of the female. The peritreme is hooked distally. The proximal pair of duplex setae are distad of the proximal tactile setae; there are no empodial spurs. Empodium I of the male consists of short tridentate appendages without a spur; the other empodia are normal. The aedeagal knob is small, about twice as wide as the stem.

Tetranychus polys has been collected in southern Arizona on *Asclepias erosa* Torr., *Atriplex canescens* (Pursh) Nutt., *Atriplex polycarpa* (Torr.), and *Pluchea sericea* (Nutt.).

Tetranychus (*Armencyhus*) Wainstein

Tetranychus (*Armenychus*) Wainstein, 1960: 149.

Type. *Tetranychus armeniaca* Bagdasarian.

The hysterosomal striae are entirely transverse. Two species of this subgenus occur in Arizona, *Tetranychus pacificus* and *T. mcdanieli*.

Tetranychus (*Armenychus*) *pacificus* McGregor

Tetranychus pacificus McGregor, 1919: 657; Pritchard and Baker, 1955: 388.

The striae of the female are transverse over the entire dorsum of the hysterosoma; the proximal pair of duplex setae are distad of the proximal tactile setae; the peritreme is hooked distally; and there is no empodial spur. Females are greenish and spotted. The empodium of tarsus I of the male consists of short tridentate appendages with a dorsomedial spur; empodium II is normal but bears a short dorsomedian spur; empodia III and IV are normal and without dorsomedial spurs. The aedeagus is directed slightly dorsad with anterior and strong posterior angulations, the posterior angulation ending well beyond the level of the bend of aedeagus forming an obtuse angle.

Tetranychus pacificus has been collected on *Ceanothus fendleri* Gray and *Rhamnus betulaefolia* Greene, in Arizona.

Tetranychus (*Armenychus*) *mcdanieli* McGregor

Tetranychus mcdanieli McGregor, 1931: 193; Tuttle and Baker, 1964: 41.

This species is similar to *Tetranychus pacificus*, differing only in the angulation of the aedeagus. In *T. mcdanieli* the dorsally directed bend of the aedeagus is sigmoid, the distal end being directed caudad at about 45° angle and terminating at a level of the posterior margin of the bend. There usually is a small, obtuse anterior angulation.

Tetranychus mcdanieli has been taken on *Morus* sp. and *Thermopsis pinetorum* Greene, in Arizona.

131

REFERENCES CITED

Athias-Henriot, C. 1961. Nouveaux acariens phytophages d'Algerie (Actinotrichida, Tetranychoidea: Tetranychidae, Linotetranidae). Ann. de l'Ecole National d'Agriculture d'Alger, 3(3): 1-10.

Anwarullah, M. 1966. *Porcupinychus abutilioni* (Acarina: Tetranychidae, a new mite from Pakistan. Can. Ent., 98(1): 71-75.

Bagdasarian, A. T. 1951. Contributions to the fauna of spider mites (fam. Tetranychidae) of Yerevan and its environs. Akad. Nauk Armianskoi S.S.R. Izv. Biol. i Sel'slokhoz. Nauk, 4(4): 368-374. (In Russian)

_____. 1954. New species of tetranychid mites from Armenia. Akad. Nauk Armianskoi S.S.R. Dok., 18(2): 51-56. (In Russian)

_____. 1957. Fauna Armianskoi S.S.R. Tetranychid mites (superfamily Tetranychoidea). Akad. Nauk. Armianskoi S.S.R. Zool. Inst., pp. 1-163. (In Russian)

Baker, E. W. and A. E. Pritchard. (1962) 1963. Arañas rojas de America Central. Rev. Soc. Mex. Hist. Nat., 23: 309-340, illus.

Banks, N. 1900. The red spiders of the United States (*Tetranychus* and *Stigmaeus*). U. S. Dept. Agr. Div. Ent. Tech. Ser., 8: 65-67.

_____. 1912. New American mites. Proc. Ent. Soc. Wash., 14: 96-99.

_____. 1917. New mites, mostly economic (Arach. Acar.), Ent. News, 28: 193-199.

Beer, R. E. and D. S. Lang. 1958. The Tetranychidae of Mexico. Univ. Kans. Sci. Bull., 38 (2) 15: 1231-1259.

Berlese, A. 1886. Acari dannosi alle piante coltivati, 31 pp. Padova.

_____. 1913. Acarotheca Italica, 221 pp. Firenze.

Boisduval, A. 1867. Essai sur l'entomologie Horticole, 648 pp. Paris.

Boudreaux, H. B. 1956. Revision of the two-spotted spider mite (Acarina, Tetranychidae) complex, *Tetranychus telarius* (Linnaeus). Ann. Ent. Soc. Amer., 49: 43-48.

_____ and G. Dosse. 1963. Concerning the names of some common spider mites. Advances in acarology. Cornell Univ. Press, Ithaca, New York, pp. 350-364.

Canestrini, G. 1889. Prospetto dell'acarofauna Italiana. Famiglia dei Tetranychini. Atti Reale Ist. Veneto Sci. Let. Art (ser. 6), 7: 491-537. Also published separately, 1890, as Prospetto dell'acarofauna Italiana, 4: 427-540.

DeLeon, D. 1957. Two new *Eotetranychus* and a new *Oligonychus* from southern Florida (Acarina: Tetranychidae). Fla. Entom., 40(3): 111.

Donnadieu, A. L. 1875. Recherches pour servir à l'histoire des Tétranyques. Ann. Soc. Lyon, 12: 1-134. Also published in 1876, Ann. Soc. Linn. Lyon (n. ser.), 22(1875): 34-163.

Dosse, G. and H. B. Boudreaux, 1963. Some spider mite taxonomy involving genetics and morphology. Advances in acarology, Cornell Univ. Press, Ithaca, New York, pp. 343-349.

Dufour, L. 1832. Description et figure du *Tetranychus linteraricus*, Arachnide nouvelle de la tribu des Acarides. Ann. Sci. Nat. Paris, 25: 276.

Ehara, S. 1956. Some spider mites of northern Japan. J. Fac. Sci. Ser. VI Zool., 12(3): 244-258.

_____. 1956. Notes on some tetranychid mites of Japan. Jap. J. Appl. Zool., 21(4): 139-147.

_____. 1956. Tetranychoid mites of mulberry in Japan. J. Fac. Sci. Ser. VI Zool., 12(4): 499-510.

————. 1960. On some Japanese tetranychid mites of economic importance. Jap. J. Appl. Zool., 4(4): 234-241.

————. 1962. Tetranychoid mites of conifers in Hokkaido. J. Fac. Sci. Ser. VI Zool., 15(1): 157-175.

————. 1964. The tetranychoid mites of Japan. Acarologia 6: 409-414.

Ewing, H. E. 1909. New species of Acarina. Trans. Amer. Ento. Soc., 35: 401-415.

————. 1913. The taxonomic value of characters of the male genital armature in the genus *Tetranychus* Dufour. Ann. Ent. Soc. Amer., 6: 453-460.

————. 1926. Two new spider mites (Tetranychidae) from Death Valley, California (Acarina). Ent. News 37: 142-143.

Jacobi, A. 1905. Eine Spinnmilbe (*Tetranychus ununguis* n. sp.,) als Koniferenschädling. Naturw. Zts. Land. Forst., 3: 239-247.

Koch, C. L. 1836. Deutsche. Crustacea, Myriopoda. Arachinida, Fasc. 1: 8.

Linnaeus, C. 1758. Systema naturae, 1 (10th ed.), 824 pp. Stockholm.

McGregor, E. A. 1916. The citrus mite named and described for the first time. Ann. Ent. Soc. Amer., 9: 284-290.

————. 1919. The red spiders of America and a few European species likely to be introduced. Proc. U. S. Natl. Mus., 56: 641-679.

————. 1928. Descriptions of two new species of spinning mites. Proc. Ent. Soc. Wash., 30 (1): 11-14, 15.

————. 1931. A new spinning mite attacking raspberry in Michigan. Proc. Ent. Soc. Wash., 33(8): 193-195.

————. 1934. A new spinning mite on citrus at Yuma, Arizona. Proc. Ent. Soc. Wash., 36(8-9): 256-259.

————. 1936. Two spinning mites attacking incense cedar in California. Ann. Ent. Soc. Amer., 29(4): 770-775.

————. 1939. The specific identity of the American date mite; description of two new species of *Paratetranychus*. Proc. Ent. Soc. Wash., 4(9): 247-256.

————. 1941. A new spinning mite attacking strawberry on the mid-Atlantic Coast. Proc. Ent. Soc. Wash., 43(2): 26-28.

————. 1943. A new spider mite on citrus in southern California (Acarina: Tetranynidae) Proc. Ent. Soc. Wash., 45(5): 127-129.

————. 1945. A new genus and species of tetranychid mite from California. Proc. Ent. Soc. Wash., 47: 100-102.

————. 1950. Mites of the family Tetranychidae. Amer. Midl. Nat., 44(2): 257-420.

————. 1952. A new spider mite (Acarina: Tetranychidae). Proc. Ent. Soc. Wash., 54(3): 142-144.

Meyer, M. K. P. and P. A. J. Ryke. 1959. A revision of the spider mites (Acarina: Tetranychidae) of South Africa with descriptions of a new genus and new species. Jour. Ent. Soc. S. Afr., 22(2): 330-366.

————. 1965. Personal communication.

Miller, L. W. 1966. The tetranychid mites of Tasmania. The papers and proceedings of the Royal Society of Tasmania, 100: 53-66.

Morgan, C. V. G. and N. H. Anderson. 1957. *Bryobia arborea* n. sp. and morphological characters distinguishing it from *B. praetiosa* Koch (Acarina: Tetranychidae). Can. Ent., 89(11): 485-490.

Müller, O. F. 1776. Zoologiae Danicae Prodromus, 282 pp. Copenhagen.

Murray, A. 1877. Economic Entomology, Aptera, 433 pp., Chapman and Hall, London.

Oudemans, A. C. 1931. Acarologische Aanteckeningen CVI. Ent. Ber., 8(177): 189-204.

_____. 1931. Acarologische Aanteckeningen CVII. Ent. Ber., 8(178): 221-236.

Pritchard, A. E. and E. W. Baker. 1955. A revision of the spider mite family Tetranychidae. Mem. Pac. Coast Ent. Soc., 2: 1-472.

Reck, G. F. 1948. Contributions to the fauna of spider mites. (Tetranychidae; Acari) in Georgia. Akad. Nauk Gruzinskoi S.S.R. Inst. Zool. Trudy, 8: 175-185. (In Russian)

_____. 1950. Materiali k faune pautinnikh kleshchei Gruzi (Tetranychidae; Acarina). Akad. Nauk Gruzinskoi S.S.R. Inst. Zool. Trudy, 9: 117-134. (In Russian)

_____. 1952. O nekotorikh osnovakh klassifikatsii tetranikovikh kleshchei. Soobsh. Akad. Nauk Gruzinskoi S.S.R. 13(7): 420-425. (In Russian)

_____. 1959. Identification of tetranychid rules. Fauna Zakaukazia, Akad Nauk Gruzinskoi S.S.R. 1: 154. (In Russian)

_____ and A. T. Bagdasarian. 1948. A new genus of the fam. Tetranychidae (Acari) from Armenia. Akad. Nauk Armianskoi S.S.R. Dok., 9(4): 183-186. (In Russian)

Riley, C. V. 1890. The six-spotted mite of the orange (*Tetranychus 6-maculatus* n. sp.). Insect Life, 2: 225-226.

Rimando, L. 1966. A new subfamily of spider mites with the description of a new genus and two species (Acarina: Tetranychidae: Aponychinae). The Philippine Agriculturist, 50: 105-113.

Ryke, P. A. J. and M. K. P. Meyer. 1960. The parasitic and predacious mite fauna (Acarina) associated with *Acacia karroo* Hayne in the Western Transvaal. Libro homenaje al Dr. Eduardo Caballero y Caballero. Jubileo 1930-1960. Inst. Polit. Nac., Escuela Nac. Cien. Biol. Mex., pp. 559-569.

Scheuten, A. 1857. Einiges über Milben. Arch. Naturg., 23(1): 104-112.

Summers, F. 1953. *Bryobia curiosa,* a new species, from the Mohave Desert in California (Acarina: Tetranychidae). Ann. Ent. Soc. Amer., 46(2): 290-292.

Trägårdh, I. 1915. Bidrag till kännedomen om spinnvalstren (*Tetranychus* Duf.). Medd. Centralanst. Försöds. Jordbr., 109 (Ent. Ard. 20) 1-60; and Stockholm Landtbr.-Akad. Handl. 54: 259-310.

Tuttle, D. M. and E. W. Baker. 1964. The spider mites of Arizona. Univ. Ariz. Tech. Bull., 158: 1-41.

Ugarov, A. A. and V. V. Nikolskii. 1937. K sistematike sredneazyatakogs pautinnovo kleshchika. In Voprocy Zachchity Khlopchatnika. Trudy Sredneaziatskoi Stantsii Zashchaty Rastennii, pp. 26-64. (In Russian)

Wainstein, B. A. 1956. Material on the fauna of tetranychid mites of Kazakhstan. Trudy Respublik. Stentsii Zashch. Rast. Kazfilial Vaskhnil, 3: 70-83. (In Russian)

_____. 1960. Tetranychoid mites of Kazakhstan (with revision of the family). Kazakh. Akad. Sel'sk. Nauk. Nauch.-Issled. Inst. Zash. Rast. Trudy, 5: 1-276. (In Russian)

_____. 1961. On the systematic position of two species of Tetranychidae mites (Acariformes) with a description of two new genera and a tribe. Akad. Nauk. S.S.R. Zool. Zhur., 40(4): 606-608. (In Russian)

Womersley, H. 1940. Studies in Australian Acarina Tetranychidae and Trichadenidae. Trans. Roy. Soc. S. Austral., 64(2): 233-265.

Yokoyama, K. 1929. New textbook of sericultural insect pests (Saishin Nippon Sangyo Gaichu Zensho). 569 pp. (In Japanese)

INDEX OF SPIDER MITES

137

138

139

INDEX TO GENERA OF HOST PLANTS

142